SCOTT W. STARRATT

Rare and violent events through geological time are the theme of this readable and thought-provoking view of the Earth's history. The evidence for episodic events and rare 'catastrophic' happenings have been gleaned from the geological record during the author's travels all over the world. Such events are shown to dominate over the gradual and continuous processes that we see in the record of the history of the Earth. From hurricanes to episodic evolution, from colliding continents to asteroid impacts – the importance of these events are presented with many illustrations, both pictorial and anecdotal.

Jargon free and entertaining – the ideas presented in this book will stimulate the student, provoke the professional and provide an enjoyable read for all.

The New Catastrophism

I make no apology for emulating the great Charles Lyell in having the Roman pillars at Pozzuoli near Naples as the frontispiece of my book. Lyell used them as the frontispiece of his *Principles of Geology* to be the great symbol of uniformitarianism, with the borings of marine animals in the pillars confirming changes in sea-level during historic times. Rudwick (1990) called them a 'secular icon' as they were printed over and over again and finally gilded in the repeated editions of Lyell's monumental work.

I use them, however, to symbolize *The New Catastrophism*, since they do not indicate a gradual change in sea-level, as implied by Lyell. There are no borings lower down in the masonry and the sea-level change must have been very rapid (Ager 1989). Therefore, though they certainly confirm Lyell's great new ideas about present-day processes, they also indicate that those changes can be very sudden. The rise and fall of the sea here is related to local episodic volcanicity which is, in my sense, catastrophic.

The New Catastrophism
The importance of the rare event in geological history

DEREK AGER

Emeritus Professor of Geology
University College of Swansea

CAMBRIDGE
UNIVERSITY PRESS

Published by the Press Syndicate of the University of Cambridge
The Pitt Building, Trumpington Street, Cambridge CB2 1RP
40 West 20th Street, New York, NY 10011-4211, USA
10 Stamford Road, Oakleigh, Melbourne 3166, Australia

© Cambridge University Press 1993

First published 1993

Printed in Great Britain at the University Press, Cambridge

A catalogue record for this book is available from the British Library

Library of Congress cataloguing in publication data available

ISBN 0 521 42019 9 hardback

RL

I dedicate this book to the memory of the splendid Department of Geology at the University College of Swansea, founded in 1920, and closed as a result of Government policies in 1990.

This book is also dedicated with gratitude to all the medical staff who kept me alive long enough to finish writing it.

Contents

Preface

This book arose largely from two visits to Banff in the Canadian Rockies as one of the main speakers at the annual National Conference on Earth Science, jointly sponsored by the Canadian Society of Petroleum Geologists and the University of Alberta. The first of the conferences in which I was involved, in 1968, was on the topic 'Fossils, Ecology and Environments of Carbonate Rocks'. The second, in 1987, was on 'Clastic Shorelines – Processes to Preservation'. In each case my function was to present four lectures relating modern processes to the ancient record (and sometimes vice versa). Several of the chapters in this book therefore represent the crystallization of my ideas in preparation for my lectures at Banff. They also represent the development of some of the ideas summarized in my book *The Nature of the Stratigraphical Record* (Ager 1973, 1981a & 1993). They extend beyond the limits of that book, however, to cover many fields besides stratigraphy and to take note of developments in the intervening years in almost all aspects of the great science of geology. My intention is to show that catastrophism (in my sense) or at least episodicity, is apparent in everything from storm deposits to evolution, from plate tectonics to volcanism, from earthquakes to extra-terrestrial impacts.

I should, perhaps, say something about the title of this book. Just as politicians rewrite human history, so geologists rewrite earth history. For a century and a half the geological world has been dominated, one might even say brain-washed, by the gradualistic uniformitarianism of Charles Lyell. Any suggestion of 'catastrophic' events has been rejected as old-fashioned, unscientific and even laughable. This is partly due to the extremism of some of Cuvier's followers, though not of Cuvier himself.

On that side too were the obviously untenable views of bible-oriented fanatics, obsessed with myths such as Noah's flood, and of classicists thinking of Nemesis. That is why I think it necessary to include the following 'disclaimer': **in view of the misuse that my words have been put to in the past, I wish to say that nothing in this book should be taken out of context and thought in any way to support the views of the 'creationists' (who I refuse to call 'scientific').**

My thesis is that in all branches of geology there has been a return to ideas of rare violent happenings and episodicity. So the past, as now interpreted by many geologists, is not what it used to be. It has certainly changed a great deal from what I learned about it in those far-off days when I was a student.

Derek Ager,* Swansea, 1992

* Derek Ager died peacefully in Swansea during the final stages of production of this book.

Acknowledgements

I would like to thank many people around the world for implanting or modifying my views on so many different topics, but it would be impossible and invidious to name them. What is more, it might seem to imply that they necessarily agreed with what I say. I hope that I have included in the text and bibliography the names of those who were clearly responsible for particular ideas and information.

I must, however, take this opportunity to offer especially warm thanks to my wife, Renée. I have never forgiven myself for taking her so much for granted that I omitted thanking her in my last book *The Geology of Europe*. For many years she has endured uncomfortable travels around Europe, North Africa and North America, living in tents, motor-caravans or simple hotels and motels. She has persevered through heat, cold, insects and my impatience to be over the next horizon. Since then our travels have taken us again to North America, to East Africa, the Far East and to the South Pacific. On all our private travels she has made all the arrangements in her very efficient way. She has also, in the last few years, had to do all the driving as I am not now allowed to drive (for medical reasons, not alcohol!). So I do thank her now most sincerely for all that she has done and suffered as a result of my strange habits, interests and what she calls my bad temper. She also nobly read part of the manuscript to draw my attention to omissions, repetitions and so on, but she is in no way responsible for the frailties that remain. The index was only possible with her help. I must apologize to her too, for implanting a wanderlust in our roaming children: our single-handed sailor daughter Kitty and our Third World water engineer son Martin. I must thank them too for assuming, when growing up, that this was a normal way of life. They also provided much practical help and honest criticism.

Finally I would like to thank Catherine Flack and Douglas Palmer of Cambridge University Press for all the trouble they have taken in bringing this book to reality.

Introduction

The only advantages I have noted about getting older are firstly that I am not afraid of women any more and secondly that I am no longer afraid of making a fool of myself. The second of these foolhardy expressions of confidence may well become obvious in this book. My specialism, and these days every scientist must have one, is one small corner of the vast field of palaeontology. In recent years, however, this has not stopped me extending into regional geology and into the fundamental nature of the geological record and what it tells us about the events of the past. I am certainly not a petrologist, a structural geologist or a sedimentologist, but I can claim to have seen a lot of rocks in many countries (57 so far, not counting transit lounges). My motto therefore, as I have proclaimed before, is that of the great Joachim Barrande: 'c'est ce que j'ai vu' (it is what I have seen) or, as they are supposed to say in Missouri: 'show me!'. In some cases, it is the view of an onlooker, who often sees more of a battle than those actually taking part.

I would also like to include at this point what I may call a 'non-precept', borrowed from Alan Lord (of University College London and acquired by him at an early age):

To make a name for learning,
When other routes are barred,
Take something very simple –
And make it very hard.

I am not in favour of making things hard. Geology, in my opinion, can be written about very simply in the language of Shakespeare. What could be simpler and yet more full of meaning than Hamlet's 'To be or not to be'? Or my favourite quotation from the great bard, when Iras says to Cleopatra, with a hint of catastrophism: 'Finish, good lady; the bright day is done, And we are for the dark'. I do not believe in C. P. Snow's 'Two Cultures', so I hope the duller and more conventional scientists will excuse my various references to 'cultural' matters, ranging from Chinese philosophy to Japanese music and from the Bible to Flanders and Swann.

I also hate jargon, which usually adds nothing to our understanding.

Thus 'model', in my view, is just a word for people who cannot spell 'hypothesis'. When I was asked to talk about 'clastic shoreline models', I immediately thought of scantily dressed young women lying on sandy beaches. My sedimentological jargon is probably three generations out of date (*calcarenite* was a wicked new word when I was a student). I hope that there will be little or no jargon in this book.

Basically I am concerned with what happened in geological history and this can be expressed quite simply in everyday English. Gibbon said that history was 'little more than the register of the crimes, follies and misfortunes of mankind'. We can hardly make moral or intellectual judgments about earth history, so the first two do not apply here. We can say, however, that the geological record is very much a matter of misfortunes, whether they are the extinctions of life or the collisions of continents.

I also remember the words of another clear-thinking man, George Bernard Shaw, speaking through the mouth of one of his characters in *The Devil's Disciple*. When asked what history will say of certain events in the American War of Independence, the British officer replies: 'History, Sir, will tell lies as usual'. The geological record is constantly lying to us. It pretends to tell us the whole truth, when it is only telling us a very small part of it. It is 'economical with the truth' as was said at a recent enquiry into British bureaucracy. Sometimes the geological record conceals or confuses the truth by diagenesis or metamorphism, like an unnamed politician wiping out the record on an incriminating tape. Very often it removes large sections of the record, like that same politician removing cards from a filing cabinet.

I would call all this 'stratigraphy', but the Americans, with sickening logic, have restricted that term to the study of stratified rocks. It was in my sense of the word that I wrote my book on *The Nature of the Stratigraphical Record* (1973, 1981a & 1993), but it so happens that most of that book was concerned with strata of various kinds. In this book I am scattering my thoughts more widely, I hope not figuratively, on stony ground. I must emphasize that I am concerned with the whole history of the earth and its life and in particular with the dangerous doctrine of uniformitarianism.

Perhaps the only major aspect of geology I do not consider in this book is that of tectonism. This is partly from fear of my own ignorance and partly because, in a way, that subject seems to have gone in the opposite direction to other branches of geological thought. Referring back to those days when I was a student, we then had no doubt that orogenies were rare events, widely separated in time. It then became apparent that there were many phases of mountain building movements, each of them of short duration and therefore what might be called episodic catastrophes. I note that the eminent tectonician Forese-Carlo Wezel thinks very similarly (1988 & 1992).

Rudolf Trümpy, in his address as President of the International Union of Geological Sciences (1980) summed up the previous 10 years as follows:

We have come a long way from the positivist, ruddy-faced uniformitarianism of ten years ago.

Geologists are beginning to realise that even improbable events become probable during a sufficiently long time span. Catastrophism is probably the wrong word, but episodicity and periodicity . . . loom large in the minds of today's geologists.

Though I agree with him, I trust that great Swiss geologist will excuse me if I prefer to retain the word 'catastrophism', if only because it is more evocative and represents more clearly the antithesis of Lyell's boring doctrine.

Similarly the eminent German palaeontologist, Dolf Seilacher wrote of the fossil record (in Einsale & Seilacher 1982):

The present generation of earth scientists has become aware that the history of the biosphere is not only one of gradual and stately changes but that it is accentuated by events of various kinds and degrees, most of which are so rare that they refute a uniformitarian approach.

Charles Lyell had a 'steady state' view of the Earth and its life. He conceded that many species had become extinct, but could not accept the idea that the Earth was once wholly inhabited by very simple forms of life. He even thought that all processes, including life, were cyclic, and the dinosaurs might reappear (discussed in Gould 1987, p. 103; Buffetaut 1987, p. 91). Indeed, once the large monsters of the past were discovered, a persistent theme in popular literature was that of supposedly extinct species surviving to the present day, either underground, as in Jules Verne's *Journey to the Centre of the Earth* or on remote table mountains as in Conan Doyle's *The Lost World* or brought back by genetic engineering, using chiefly their blood in blood-sucking insects preserved in amber, as in Michael Crichton's recent book *Jurassic Park* (1991).* The idea is also still with us in films with hopeless time-scales, such as *One Million Years BC* or in the various versions of *King Kong* and its derivatives. Even people as eminent as Thomas Jefferson (second president of the USA) expected supposedly extinct species to be found still alive in the remoter parts of the earth (see Buffetaut 1987).

It must be appreciated, of course, that geology as a science is less than 200 years old. When Lyell and his contemporaries were working, the Pleistocene (and other) glaciations were unknown; the Lower Palaeozoic

---

* Unfortunately most of the dinosaurs in that book are Cretaceous in age!

and certainly the Precambrian rocks were thought to be too complex for understanding and much of the world was geologically unexplored.

One must constantly ask oneself 'Is the present a long enough key to unlock the secrets of the past?' I shall discuss this further in Chapter 12. Even within the brief life of mankind (with 99% of it in the 'Stone Age') there were great geological events that are not recorded in our histories. There was active volcanicity in many parts of Europe, such as Bohemia and the Auvergne, where it is now unknown. There was the tremendous explosion of Santorini in the Aegean (perhaps the loudest noise ever heard by humans). There was a similar even greater explosion in what is now Lake Taupo in the northern island of New Zealand. There was that great torrent of lava pouring into the Grand Canyon and the unimaginable flood of water sweeping across the 'barren scablands' of the north-west United States, cutting canyons scores of metres deep in hard basalt. There were probably earthquakes and storm surges vastly more powerful than any that have been recorded on scientific instruments. All these things will be discussed in later chapters.

Charles Lyell the arch-priest of uniformitarianism of course knew something about present-day violent happenings such as eruptions and earthquakes, but was not in a position to apply that knowledge properly to the past.

To many it may seem that the arguments between 'catastrophism' and 'uniformitarianism' are long past, with the latter victorious. But on the principle that one cannot slander a dead man, it is easy to think that our predecessors were fools. I was struck by a remark by the Dutch geologist Hooykas (1984) that 'we have to keep in mind that our ancestors' eyes were as good as ours'. On those grounds I think that the 'catastrophist' Georges Cuvier was a better geologist than the 'uniformitarian' Charles Lyell and my first chapter is devoted to a defence of that great Frenchman. Whilst Lyell generalized and theorized, Cuvier recorded what was actually there, bed by bed, and in doing so noted the frequent changes in environment and the gaps in the record. He was also probably the first to recognize that species had become extinct during the troubled history of the Earth.

Darwin (1859) thought that 'long periods of time have left virtually no geological record, and that slow extermination might well have taken place during these unrecorded periods'. But due to our subconscious wish for the complete story, we have tended to ignore this possibility. I will argue that many of our present-day environments and their life, are not being preserved for the geologist of the future.

Much of this kind of thought is a matter of the resolving power of our geological eye-glass. Thus, as J. Maynard Smith pointed out (1981). 'the idea of new species arising within 50 000 years of each other seems sudden to a palaeontologist, but gradual to a geneticist'. However, we are here concerned not only with life, but with the rocks which record the whole history of the Earth.

It seems to me that there are four fundamental factors in the formation of the geological record:

1 The passage of time. There's not much we can do about this, except to note that there was vastly more of it than we tend to appreciate.

2 Everyday processes, which go on all the time, like T. S. Eliot's summation of life: 'Birth, and copulation, and death. That's all the facts when you come to brass tacks . . .'. Or to put it in geological terms: erosion/deposition, intrusion/extrusion, tectonism/metamorphism. These seem to be the main preoccupations, I might say fixations, of geology.

3 Things happen, that is to say the unusual happenings, the rare events, which I think are far more important than is generally appreciated.

4 Some of the record, probably only a very little of it, is preserved and this process of preservation is just as important, probably more important in fact, than the original process of formation (see Chapter 2).

It started, when I was a student, with turbidity currents. Instead of sedimentation keeping pace with subsidence, we were allowed to think of troughs which could be rapidly filled with sudden rushes of turbid water. There followed the recognition of numerous sudden events, ranging from storms and submarine landslips, episodic evolution and extinction to extra-terrestrial bodies impacting on the Earth.

These ideas are not new and everyone does not agree, but in general it may be said (again recalling T. S. Eliot) that bangs have replaced whimpers and the geological record has become much more exciting than it was thought to be.

In a phrase that has often been quoted since, I have summed up geological history as being like the life of a soldier: 'Long periods of boredom and short periods of terror' (Ager 1973, 1981a, 1993). So that is the message of this book too. This is not the old-fashioned catastrophism of Noah's flood and huge conflagrations. I do not think the bible-oriented fundamentalists are worth honouring with an answer to their nonsense. No scientist could be content with one very ancient reference of doubtful authorship. For different reasons I do not go along completely with the modern fashion of cosmic collisions or with dark stars causing periodic mass extinctions. To me *The New Catastrophism* is mainly a matter of periodic rare events causing local disasters. Just as Gibbon documented the decline and fall of the Roman Empire, so we may record the rise and fall of the trilobites or the periodic spreading and retreating of marine transgressions across our continents.

I should make reference here to four other volumes in this field. There is that by Clube & Napier (1982), there is the volume edited by Berggren & van Couvering (1984) in which I participated, also Hsu (1986) and that by Albritton (1989). But these all consider specific issues and events, whereas I am concerned with every facet of earth history from storms to sexuality, from earthquakes to asteroids.

Special note: I am avoiding the term 'billion' in this book because it means different things to different people.

1

A much misunderstood man

Georges Cuvier was, in the estimation of Stephen Gould '... perhaps the finest intellect in nineteenth century science ...' (1987, p. 113). He has also been called (by Jane Gregory 1988) 'an unsung champion of science'. Certainly he has always been a hero of mine, so much so that he became a joke in my department (at the University College of Wales, Swansea) and one year my colleague who managed our examinations inserted in the rubric at the beginning of the last paper (which consisted of 100 short questions): 'The answer to one of these questions is Baron Cuvier'.

Cuvier was a remarkable man. He was the 'father' of vertebrate palae-ontology, he began the study of comparative anatomy and of functional morphology (that is to say he recognized the close relationship between the detailed anatomy of an animal and its mode of life). He was the first man to observe clearly that species became extinct and, in my opinion, he was the first man to understand the geological record as it really is. After many violent and, one might say, catastrophic changes in the history of France at the end of the eighteenth and beginning of the nineteenth centuries, he became a baron and nearly became president of the republic. He also earned himself a statue outside the Museum of Mankind in London. Yet he became a figure of fun among geologists and was falsely blamed for all the ridiculous excesses of nineteenth century 'catas-trophism'.

By all accounts he was a rather dull man, but with much effort and practice he became a brilliant lecturer and writer. After the French Revol-ution he was one of a team given the task of reorganizing French science. He was successful (as all eminent scientists need to be) in persuading his government to provide funds for research rather than for arms and he was personally responsible for encouraging a great public interest in natural history.

Unfortunately he does not appear to have been a very nice man; thus he poked fun at Lamarck's blindness and at the same time rejected the latter's pioneering ideas about evolution. His reasons for opposing the concept of evolution were simple. Thus he could not see how the giant ground-sloths of the Pleistocene could have given rise to the much smaller sloths of today. Similarly, the many mammalian fossils he excavated from the Early Tertiary deposits of the Paris Basin (including the gypsiferrous beds that were used to make 'Plaster of Paris') could not, at that time, be seen as possible ancestors of living forms. He could see how they functioned, whether they were carnivores or herbivores, browsers or grazers, swimmers or runners, but he could not connect them, through the great gap of later Tertiary times, with the animals of today. He did not know about the faunas of the 25 million or more years that separated his Palaeogene forms from those already known in the Pleistocene. He described some 300 new species. Many of his battles on behalf of science are sickeningly familiar to us in Britain today. Thus he reminded the authorities of the value of museums and exhibitions which 'speak ceaselessly to the eye, and inspire a taste for science in young people'. He campaigned continuously for the education of the young in science and, although his own research might be regarded as purely academic, he understood that science could be useful to the people and the state. He wrote 'The present era will be reproached if we do not conserve for the future these sources of so many advantages'. Those same arguments are very familiar to many of us today, and yet so little heeded by the government, even when the previous prime minister was herself said to be a scientist!

Nevertheless, in spite of his obvious brilliance and foresightedness, Cuvier became the 'villain of the piece' to most geologists, for proclaiming the 'old-fashioned' ideas of catastrophism in conflict with the 'modern' ideas of uniformitarianism, championed by the 'real scientist' Charles Lyell. But both great men, interpreted in this way, are what Gould rightly calls 'cardboard figures'.

Lyell was a great theorizer. He looked at the modern world and the physical processes that were affecting it. He then deduced that they had functioned in the same way in the geological past. He was what has been called a 'substantive uniformitarian', that is to say that he presumed that these processes had always operated in the same way and at the same rates as they do today. After the publication of the second volume of his great *Principles of Geology*, G. P. Scrope (who had studied the volcanoes of central France) wrote in 1832 to congratulate him in the following terms: 'It is a great treat to have taught our section-hunting quarry men that two thick volumes may be written on geology without once using the word "stratum"'. That sounds to me suspiciously like 'my mind is made up, don't confuse me with facts!' That could never be said of Cuvier or that much humbler man William Smith in England. They were great pragmatists and were wholly concerned with strata and what they contained. It was said of Smith, who first demonstated the succession of strata and their fossils along the sides of his canals and in mines, that he had 'opened the book of earth history'.

Cuvier studied the stratigraphical succession in the Paris Basin and recorded stratum by stratum what he saw. In 1822, together with another palaeontologist Brongniart (who described the fossil plants), Cuvier carefully recorded many sections in the Eocene and Oligocene strata around Paris. He did not go in for grand theories about earth history, he simply placed on record what he saw. He saw repeated changes in environment and repeated changes in the faunas in the quarries around Paris. He saw freshwater limestones and marine sands, he saw shell-banks and beds without fossils. There were terrestrial deposits too, of course, such as are seen in the huge quarry at Cormeilles near Paris, with the land mammals that he studied so carefully, and there is the famous *Calcaire grossier*, packed with exotic-looking shells and well seen in the walls along the Seine in the French capital. Cuvier's general section for the Eocene of the Paris Basin is shown in Figure 1.1. This may be translated (from the top down) as follows:

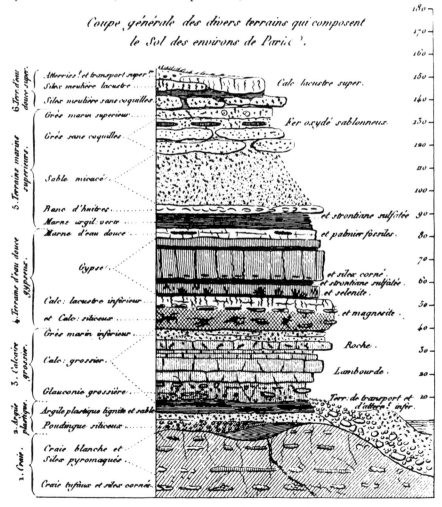

Figure 1.1 General succession in the Paris Basin (France) from Cuvier & Brongniart (1822). Translation in text.

General section of the various beds which form the ground beneath the sur-
roundings of Paris

    6. Upper freshwater beds
        Upper lacustrine limestone
        Upper residual and
           transported material
        Ground lacustrine flint
        Ground flint without shells
    5. Upper marine beds
        Upper marine sandstone   Sandy iron oxide
        Sandstone without shells
        Micaceous sands
        Bed of oysters
        Green marly clay   With strontium sulphate
    4. Gypsiferous freshwater beds
        Freshwater marl   With fossil palms
        Gypsum   With 'horned' flints, strontium
                            sulphate and selenite

        Lower lacustrine limestone
           and siliceous limestone
    3. *Calcaire grossier* (literally
          common or thick limestone)
        Lower marine sandstone
        Calcaire grossier s.s.   Rock
                           Soft, chalky stone
        Coarse glauconite   Lower transported and detrital bed
    2. Sandy, lignitic plastic clay
        Siliceous conglomerate
    1. Chalk
        White chalk with flints
        Lower chalk with 'horned'
          flints*

This seems to me a remarkably detailed work for its time and shows
real appreciation of the different environments involved. This is how the
stratigraphical record usually is, in my experience, with frequent changes
of environments, lithologies and fossils especially, as here, in shallow
water and terrestrial deposits.

As Cuvier himself said 'Life on earth has been frequently interrupted
by frightful events. Innumerable organisms have been the victims of such
catastrophes. Invading waters have swallowed up the inhabitants of dry
land. Their species have vanished for ever.' (quoted by Wendt, 1970).
Of course, Catholic France put its trust in the church, rather than in the
bible and did not go through the agonies of Protestant Britain in relating

* this would not be the 'Lower chalk' as generally understood in Europe (i.e. Cenomanian in age
  and usually lacking flints).

geological discoveries and evolution to what was regarded as the absolute truth as recorded in holy writ. So much for the false linking of Cuvier with the biblical fundamentalists. He also argued (1827) that geological changes were sudden and that the present continents were in a geological sense very new. Of particular interest is his description of the pillars which form my frontispiece, many years before they were described by Lyell: 'the example of the Temple of Serapis, near Pouzzola [sic] proves that the margins of the sea are, in many places, of such a nature as to be subject to local risings and fallings'. He went on to point out that many other Roman quays, roads etc., from Alexandria to Belgium, were not affected. This was simply because they were not affected by local volcanicity, as was (and is) the Naples area.

When he used the word 'catastrophe', for which he has been so much blamed, he was thinking (as I am) of the repeated comings and goings of the sea, which were not always 'plus ou moins graduelle' (more or less gradual) and could be very general. The land animals obviously perished when their land was flooded. He commented on the remains of land animals and plants in beds which have long disappeared 'au milieu des couches marines' (in the middle of marine beds). He did note, however, that 'lorsque la mer a quitté nos continents pour la dernière fois, ses habitans ne différaient pas beaucoup de ceux qu'elle alimente encore aujourd'hui' (when the sea finally left our continents for the last time, their inhabitants did not differ very much from those which they still support today). So, though he did not accept Lamarckism, one can only wonder what his reaction would have been to Darwin. Unfortunately he died in 1832, just 27 years before the publication of *The Origin of Species*.

2

Magnolias and marigolds; hippos and hiatuses

Magnolias and marigolds

This will be a somewhat paradoxical chapter about what is not there. It will be concerned with preservational potential, that is to say the gamble of whether fossils and rocks are preserved for the geologist of tomorrow.

A magnolia is more likely to be preserved than a marigold, an oak tree than a daisy. This is simply because the magnolia and the oak tree are woody plants, whilst the marigold and the daisy are soft and herbaceous. Is this the only reason for the woody flowering plants having a much longer fossil record than the herbaceous sort? Is this all that lies behind the botanical theory that the woody angiosperms were ancestral to the humbler genera? The most successful flowering plants today are probably the Compositae (the daisy family, including the marigold) and the Umbelliferae (the parsley family). In fact I sometimes think that it is not the triffids (of John Wyndham's science fiction novel) which are taking over the Earth. It is the umbellifers and the sycamores. But I speak from the view-point of a Briton who has seen the beautiful flowery hedgerows of his childhood disappear under the bureaucratic 'weed' killers which seem to favour the boring umbellifers, sycamores and brambles.

I have a private theory that many of the ancient plants that have survived to today are those with a built-in resistance to dangers that affect their later, weaker brethren. Thus while cow-parsley (*Chaerophyllum sylvestre*) seems to be taking over on British road-sides, the member of the ancient fern family, known as bracken (*Pteris aquilina*) with its deep, strong roots, is taking over the moors from the heather (*Erica cinerea*) and its relations, especially after frequent heath fires. Another ancient plant which is clearly able to survive severe burning is the giant redwood (*Sequoiadendron giganteum*, often called *Wellingtonia* in Britain and *Washingtonia* in the States!) Ancient fires in what is now Yosemite National Park have in places burned these trees, often thousands of years old, right through, allowing people and cars to pass right through the blackened arches left in the still very much alive trees. So

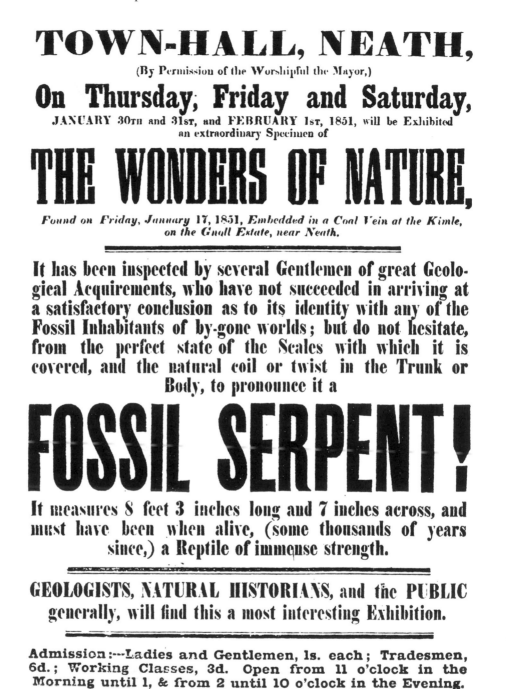

Figure 2.1   Poster displayed in Neath (South Wales) in 1851. The 'serpent' was in fact a *Stigmaria* root, such roots are commonly preserved in the local Upper Carboniferous Coal Measures. Note the prices of admission, which reflect the class consciousness of the time.

far as daisies are concerned, I was surprised to see none in Japan, where the children make chains of clover instead. So geographical distribution comes into it too.

It is not, however, only the sturdy trees and resistant ferns that have survived from ancient times. Certainly the Articulatales have a record going back to massive trees such as *Calamites* in Late Palaeozoic times, but they were already reduced to low marsh-living plants such as *Equisitites* in the Mesozoic and the modern 'horse-tails' (*Equisetum*) look far from sturdy in their usually damp habitats. They are only too easy to pull apart at their nodes, as I find when I try to pull them up in my garden. I well remember, however, returning home after my usual long (working) vacation travels in 1976, following an exceptionally dry summer (even in Wales!) to find my lawns as well as my flower-beds covered with horse-tails. Their secret, of course, like the ferns, lies in their extensive, creeping root-stock, which reminds one of *Stigmaria*, the roots of *Calamites*, which are probably the most common plant fossils in our Upper Carboniferous 'Coal Measures'.

In the middle of the last century a large *Stigmaria* was put on show in Neath, near Swansea, and called a 'Fossil Serpent' (Figure 2.1). Particularly fascinating are the charges for admission: one shilling (five new pence) for ladies and gentlemen, six pence (2½ new pence) for tradesmen and three pence (1¼ new pence) for the working classes. It does not make clear how they were distinguished in those class conscious days. Presumably the gentlemen wore top hats, the workers wore cloth caps and, in my first theory, I thought perhaps the tradesmen wore bowlers, but further research revealed that the bowler had only been invented two years before and I doubt if it had reached Neath by 1851. Fortunately those days are long past. Of course stigmarian roots, buried in the soil, stood a much better chance of preservation than 'serpents', presumably writhing about on the surface. Incidentally it has recently been pointed out by Alan Batten (in *New Scientist*) that since living 'serpents' have rudimentary legs, God's verdict on the serpent in Genesis that 'upon thy belly shalt thou go' indicates an early recognition of evolution.

Closely related to the evolution of land tetrapods is that of grass, probably the most successful group of organisms on Earth at the present day. It positively thrives on being eaten or cut. I like to remember the recipe for the perfect English lawn. One prepares the ground, one plants the seed and then one cuts it every week for 300 years. Such an organism was obviously bound to succeed, though in my own lawns there is strong competition from the Compositae, notably the common daisy (*Bellis perennis*) and the dandelion (*Taraxacum officinale*), together with what is probably a much more ancient stock, the mosses. One presumes its antiquity from the organization of the plant, not from its fossil record, which is virtually non-existent (which is hardly surprising in view of its tender form). The so-called 'club mosses', the lycopods, are very well known from the Late Palaeozoic simply because they then took the form of large forest trees. None of these modern forms, such as daisies and grass has much in the way of a fossil record, probably simply because of their non-preservability and the 'club mosses' are reduced to insignificance. I shall discuss the grasses again later (in Chapter 13), but I might mention here that their record probably goes back to the Cretaceous. It seems that they are chiefly known in the older rocks by their woody members, the bamboos, which obviously had a better chance of survival than the seemingly fragile forms on which we walk.

One important aspect of the problem of preservation or non-preservation is what is called the study of taphonomy. This is, strictly speaking, the study of burial, from the Greak word ταφος (taphos) – a tomb (as in cenotaph = empty tomb). So we must always consider the way our fossils are buried, whether they had been transported after death and whether they had changed or disappeared as the result of burial. This makes it all the more likely that the exhibit at Neath (Figure 2.1) was a root and not a reptile. I have noted the selectivity of preservation, even among single groups such as the molluscs, in the Pliocene and Pleistocene shelly sands or 'crags' of eastern England (where I found my first fossil when digging a slit trench in the expectation of an invasion in the Second World War). Some of the many molluscs there (such as my first, still cared-for *Aequipecten opercularis*) have clearly survived fossilization better than others and some fossils have been derived from elsewhere (such as rare Mesozoic brachiopods which occasionally turn up in the same deposits and which I am sent to identify). It is well known that gastropods, nearly all with aragonitic shells, are much less likely to be preserved than other fossils with calcitic shells.

I always think of the first grave-digger in *Hamlet* as a pioneer taphonomist who recorded his observations. His scientific communication to the prince was to the effect that a human corpse, if it were not rotten before he died would last eight years in the earth before it rotted or nine years in the case of a tanner whose 'hide is so tanned with his trade, that he will keep out water a great while'. In the case of poor Yorick, who had been in the ground 23 years, only the skull was recorded so we may presume selective preservation even within a single specimen. So fossilization is a highly discriminating process and I may turn now to the second part of the title of this chapter: 'hippos and hiatuses'.

Hippos and hiatuses

I once met a charming lady who turned out to be a world authority on fossil hippopotamuses (or hippopotami – before the reader criticizes me, I must record that both plurals are permissible). In my usual frivolous way I made a facetious remark about the likelihood of the average geologist finding such things. I was surprised to be told that they and their predecessors, the anthracotheres, are in fact very common so far as vertebrate fossils are concerned. In fact they are so common that many vertebrate palaeontologists almost kick them aside as many specialists in fossil invertebrates tend to disregard oysters. Then it suddenly struck me that, though I have no pretensions to be a vertebrate palaeontologist, I had actually found two myself, one in the Oligocene of the Isle of Wight in southern England, and one in the Miocene of eastern Libya.

Then I suddenly remembered that famous hippo song of Flanders and Swann about 'Mud, mud glorious mud' in which the hippos delighted as their favourite habitat. I also remembered T. S. Eliot's lines:

The broad-backed hippopotamus
Rests on his belly in the mud;

obviously therefore their preservational potential is high. They simply ask to be fossilized. They have only to die to sink into their beloved mud and to await the palaeontologist of the future.

My hero, Georges Cuvier (see Chapter 1) commented on the abundance of fossil hippos. Thus in his research on fossil bones (1820, p. 313) he wrote (my translation): 'On my first journey in Tuscany, in 1809 and 1810, I found, either in the collection at Florence or in that of the Academy of the valley of Arno at Figline, such an abundance of fossil bones of hippopotami . . . a considerable quantity I have bought from the homes of peasants.' He went on to compare hippos with other vertebrate fossils and made the point about their different modes of life.

Then I considered the other non-marine vertebrate fossils that I have found in a long career of grave-robbing. There were turtles, for example in the uppermost Jurassic of the Dorset coast in south-west England. There were crocodile scutes in various places, such as the Eocene of the Isle of Sheppey in the Thames estuary and again in the Miocene of eastern Libya. Obviously these animals also lived in an environment where they stood a very good chance of preservation.

I have only seen living hippos outside a zoo in repeated visits to East Africa, where they are common snorting away in their special environment, as are turtles and crocodiles. I saw as many as seven hippos together high up in the Tsavo River of Kenya and lower down that river I saw the same number of crocodiles. I photographed a sign which said 'Beware of crocodiles'. Three years later that sign had changed to '*Vorsicht Krokodilen*', presumably indicating a change in the tourist population.

The land fauna in East Africa, however, is vastly more abundant, notably the great columns of wildebeest or gnu that sound at night like motor-cycle rallies, the zebra, the many kinds of antelope and the sadly depleted herds of elephants (though my wife and I saw more than 40 in one place). Land mammals, however, have left a far poorer fossil record than freshwater hippos and reptiles because their skeletons just lie on the land surface and rot away. The birds too are very common and diverse in East Africa, such as the many species of kingfisher, the iridescent starlings, the ever-hovering vultures, the weaver birds, the ostriches and the beautiful red and yellow barbets. But their chances of preservation are very close to zero, not only because of their habitat but also because of the fragility of their skeletons. So we cannot expect a complete record of the land fauna of the past.

Of course, in our self-centred way, our greatest interest is in the species *Homo sapiens* and we may consider here that popular concept of early humans as 'cave people.' Nothing could illustrate better the point I wish to make in this chapter. 'Cave people' were so called because that was where their remains were found, not because they all lived in caves. That would have been impossible since caves are largely restricted to limestone country and there are no caves over the greater part of the land surface where early men (and women – excuse me) must have lived. They must have made temporary structures of branches and leaves and skins, which stood no chance whatsoever of preservation. One of the first remains of an early human to be found (1823) was in a cave in the limestone of the Gower Peninsula near my home in South Wales. This was the so-called 'red lady of Paviland' who proved later to be neither a lady nor naturally red. She, who is now a he, was evidently stained with red ochre after death, dated at 26350 ± 550 BP. 'Better red when dead' seems to have been the contemporary belief. This indicated a ceremonial placing of the body in the cave rather than that he died on his hearth-rug. My son tells me that the dead of the Dogon people in Mali are similarly placed in caves. Early hominoids were found in caves in South Africa and the so-called 'first European', *Archanthropus europeus* was found in the Petralona cave in Greece. Evidence of the wider distribution of hominoids outside caves comes from the wide scatter of stone implements, often in river gravels.

A remarkable sight is the locality at Olorgesaille in Kenya where the surface is covered with hundreds of stone axes (Figure 2.2). It seems that early people camped here for thousands of years beside a lake that has since receded over the horizon. There were no caves available and the human homes, if there were any, have long disappeared. There are no human remains that I know of, only the preservable artefacts. Hippos are preserved too, as usual, and some of the giant baboons which seem to have constituted a major part of the contemporary human cuisine. I saw no caves in the Olduvai Gorge of Tanzania, where these early people lived for so long (Ager 1992a). I also presume that there are no caves along the shore of Lake Turkana in Kenya, with its famous record of early *Homo*.

Turning to the marine habitat, the most popular place for marine invertebrates to live at the present day is the rocky shoreline, with fantastic numbers packed into small areas. Most obvious are the barnacles, limpets and mussels, as on the rocks of Rhossili at the end of the beauti-

Figure 2.2   Stone implements (approximately 20 cm long) scattered on a living surface at Olorgesaille, Kenya. Photo DVA.

ful Gower Peninsula. Such shells are, however, almost though not completely, unknown as fossils, because their habitat was not one where sedimentation normally occurred to preserve them. I must be careful about this because Darwin had enough fossil barnacles to write a monograph (1851–5) about them when he returned from his Beagle voyage.*

I should also mention that I have found fossil barnacles myself, quite abundantly, in the Pleistocene north of Wellington, New Zealand and yet again in the Miocene of eastern Libya. In both of these cases, however, the locality was close to a contemporary shoreline from which the barnacles had been derived. In Libya they were particularly large and massive specimens, able to withstand a lot of rough treatment.

Rocky shorelines and shoreline sediments are rarely preserved (see Chapter 6), though this is the marine environment most thoroughly studied by sedimentologists and marine biologists. Thus fossilization is an extremely selective matter and the same may be said of the sedimentary record. The sedimentological side of things will be discussed in later chapters, but obviously it is the same with the sedimentological record as with the organic record. A muddy lake is more likely to be preserved

* My copy had a book-plate of the Royal Artillery Institution and I like to think of those gunner officers, sitting back in their leather armchairs before going off to the Crimea, reading Darwin on fossil barnacles.

than a bare plain, the deposits of the deep are more likely to survive than those of a rocky coast.

In summary I may remind readers of that remarkable history of England *1066 and All That* by Sellar & Yeatman (1938). They start (in their 'Compulsory Preface') with the important injunction 'History is not what you thought. *It is what you can remember*' [their italics]. Similarly it may be said that earth history is not a record of what actually happened. *It is a record of what happens to have been preserved* [my italics]. In the latter case it is not preserved in human memory, but in the record of the rocks. Or in other words, as Julian Barnes says in his equally remarkable *History of the World in 10½ Chapters* (1989, p. 242): 'History isn't what happened. History is just what historians tell us'. Similarly, we do not study the history of the Earth, we only study what the rocks tell us.

In my love of 'apt alliteration's artful aid' I linked hippos with hiatuses in the title of this chapter and this seems to be the point at which to discuss the breaks in the stratigraphical record. Perhaps I have already done this *ad nauseam*, notably in Chapter 3 of my stratigraphy book (Ager 1973, 1981a & 1993) which I called 'More gaps than record', but the point cannot be emphasized too much. Nowhere in the world is the record, or even part of it, anywhere near complete. Even in the Grand Canyon of the Colorado River and the adjacent sections along the Little Colorado River, surely the finest record of geological history anywhere on Earth, there are huge breaks. Notable is the complete absence of the Upper Carboniferous (Pennsylvanian) and of the Ordovician and Silurian Systems. Devonian strata are only present in local lenses. Though thick Late Precambrian Proterozoic sediments are present in the eastern part of the Grand Canyon, farther west just part of the Cambrian rests directly on the Archaean with igneous intrusions in the latter (Figure 2.3).

We talk about such obvious breaks, but there are also gaps on a much smaller scale, which may add up to vastly more unrecorded time. Every bedding plane is, in effect, an unconformity. It may seem paradoxical, but to me the gaps probably cover most of earth history, not the dirt that happened to accumulate in the moments in between. It was during the breaks that most events probably occurred. This was when the seas and rivers moved backwards and forwards, this was when many animals and plants lived out their short lives or evolved into new forms. I shall quote again Hardy's words:

Only an old man harrowing clods,
as dynasties passed.

There is no record of the old man, there is no record of the frequently redeposited clods; there is often no record of the passing dynasties. It is the gaps in the record that really matter.

Figure 2.3 Cambrian sediments (with beds up to 1 m thick) resting unconformably on Archaean gneiss, with an acid and a basic dyke cutting the latter. Bright Angel Trail, Grand Canyon, northern Arizona, USA. Photo DVA.

I may here quote appropriately the Chinese Taoist philosopher Lao Tsu in English translation (1973, Chapter 11):

Thirty spokes share the wheel's hub;
It is the center hole that makes it useful.
Shape clay into a vessel;
It is the space within that makes it useful.
Cut doors and windows for a room;
It is the holes which make it useful.
Therefore profit comes from what is there;
Usefulness from what is not there.

The last two lines are particularly significant in a geological context. The profit, be it coal or oil or any of the other earth resources, come from what is there. The usefulness comes from the spaces, the gaps, which make sense of the vastness of geological time and without which the geological record would be of unimaginable thickness and completely incomprehensible.

Similarly in music, the gaps between the notes are just as important as the sound that is actually heard. The Japanese have a stringed instrument like a horizontal harp, called the *koto*, dating from the fifteenth century. It was played to us several times in homes while we were visiting Japan. It produces music of almost indescribable purity essentially because of the pauses between notes, the gaps, which reflect the Japanese concept of *ma*, which may be translated as 'a meaningful void'. The geological record is full of meaningful voids.

I am often irritated by people talking about 'continuous sedimentation'. Such continuity usually only exists in the minds of sedimentologists who do not bother with the palaeontological detail, but perhaps I am prejudiced. It usually means, for example, just a little bit of Ordovician followed by a little bit of Silurian, followed by a little bit of Devonian with no thought of the gaps in between.

It should be obvious from a simple consideration of the thickness of the strata plotted against the period of time involved. The average sedimentation rate on the continental shelves is something like 1 cm per 1000 years. The highest known rate is perhaps that in the Gulf of Mexico, which may reach 1 cm in 100 years. The average thickness of the record since the beginning of the Phanerozoic is about 6 km or up to 60 km if we add all the maxima together. Even the Grand Canyon does not approach that kind of figure. What was fixed in my mind as I walked down it in 1959 (Figure 2.4), and even more so as I walked up it again,

Figure 2.4   The author at the bottom of the Grand Canyon (many years ago), with Archaean gneiss in the background and the Colorado River, USA, thick with sediment, in the foreground. Photo DVA.

was that it is about a mile deep and about 17 miles wide (that's about 1.61 km and 27.37 km respectively). Sadly I couldn't do it now.

If one attempts to calculate rates of sedimentation in the past, the results are usually ludicrous. Thus we may consider my favourite ammonite zone, which is the *Pleuroceras spinatum* Zone of the Upper Pliensbachian (Lower Jurassic) on which I did my Ph.D. many years ago. If one assumes (as most authors do) that the deposition of one ammonite zone lasted about a million years, then if my zone is a metre thick (which it often is not), this means that it took 1000 years to deposit 1 mm of sediment. When it gets even thinner, as on the Dorset coast in south-west England, then the theoretical rate gets even slower and the breaks certainly become more obvious. Alternatively, one may consider the English Upper Chalk, of Late Cretaceous age. In north Norfolk (in eastern England) it totals 394.7 m in thickness (Hancock 1975); this was laid down in about 32.5 million years, which gives a rate of 12.14 m per million years. This is rather less than Jake Hancock's average of about 16.8 m from his figures for the six constituent stages. That works out at 82 372 years to deposit 1 mm of Chalk, or just under 60 000 years taking the higher figure. One must bear in mind that the Upper Chalk is essentially a pelagic deposit made of coccoliths and planktonic foraminifers, which were presumably raining down all the time. With these kinds of figures one must deduce that these deposits were not continuously deposited, but include enormous gaps, with or without contemporaneous erosion. More and more such gaps are being found everywhere. Thus McRae (1990) reported 'unusually long hiatuses' in an intermontane basin of Miocene age in Bolivia, on the basis of fine scale stratigraphical data and palaeomagnetic dating. Bolivia is one country I have not visited, but I have no doubt that such 'unusually long hiatuses' occur everywhere.

In view of all these gaps therefore, it follows that it is impossible to define stratigraphical units in terms of stratotypes. It also follows that we cannot use particular horizons to define the limits between stratigraphical units. The only solution is to use the principle of the 'golden spike', for which I have argued for years in stratigraphical committees and elsewhere (Ager 1973, 1981a, 1993, Chapter 7). The spike defines, not the boundary, but the base of the overlying unit. If the spike is driven in at the best possible locality (where gaps are least obvious), then the first grain of sediment on that spike marks the bottom of the next stratigraphical unit and the start of the next time division. It has been called the 'topless' principle in stratigraphy.

3
Modern terrestrial deposits

Catastrophic flooding

Undoubtedly the most spectacular demonstration of catastrophism in modern terrestrial sediments is provided by the flood deposits of the 'Channeled Scablands' of the north-west corner of the United States. The story of J. Harlen Bretz's battle for his remarkable theory over a period of nearly 50 years before his final success, is well known. Now other men quibble about the details. I know of no better justification for independent thought and scepticism about conventional ideas. I had the honour of meeting this remarkable catastrophist in 1959. He wrote papers about his theory from 1923 up to 1969 (see references), but finally triumphed over the geological establishment. Even the US Geological Survey (1982) yielded in the end. In 1963, the International Association for Quaternary Research met in the USA and after exploring the area sent a telegram to Bretz ending 'We are now all catastrophists'.

What then was his startling idea of catastrophic events? Bretz advanced the hypothesis in 1923 that a great flood of water had suddenly swept across the states of Montana, Idaho, Oregon and Washington through the Cascade Mountains to the Pacific, stripping off the surface sediments (Figure 3.1) hence the ugly name 'Channeled Scablands'. After thousands of years only a thin soil has developed in the area. The flood cut deep gorges or 'coulees' in solid basalt (Figure 3.2). The largest of these is the Grand Coulee, with walls up to 900 feet high (*c.* 270 m) and now sustaining the huge Grand Coulee Dam (built on a granite) which generates up to 314 000 kW for the power system of the region.

There was also a series of great waterfalls, of which the best known is Dry Falls in the Lower Grand Coulee, which are 350 feet high (107 m) and 3 miles (nearly 5 km) wide and show the position of the falls when the water was here at the end of the great flood. In my diary I say that these 'must have been a fantastic sight in Pleistocene times, with the Columbia River pouring over a fall vastly higher and wider than Niagara' (Figure 3.3). It has been estimated that the flow of water involved was 386 million cubic feet (nearly 11 million cubic metres) per second or

Figure 3.1   Typical 'Barren Scabland' country near Davenport, Washington State, USA. Photo DVA.

Figure 3.2   Coulee below Grand Coulee Dam, Washington State, USA, showing succession of basalt flows. Photo DVA.

Figure 3.3 'Dry Falls', Washington State, USA, where a vast waterfall tumbled over several basalt flows in Pleistocene times. Photo DVA.

about ten times the combined flow of all the rivers of the world (US Dept. of the Interior 1982, p. 12). For comparison, the world's largest river, the Amazon, flows at about 6 million cubic feet (nearly 170 000 cubic metres) per second. The great flood swept across the north-west states at up to 45 miles per hour (more than 72 kilometres per hour).

All this happened some 12 000 to 15 000 years ago, which I regard, for the purposes of this book, as 'modern' in geological terms. The Amerindians had probably arrived by this time. The Marmes Rock-shelter, near the spectacular Pelouse Falls in Washington has yielded human bones and human artefacts. Probably many of these early people perished in the flood, just as we read today (December 1991, as I write this) of the dreadful loss of human life in the storm surge ('tidal wave' or tsunami) that has just devastated Bangladesh. The amazing thing about Bretz's theory is that when he first proposed it, he knew of no possible source of this immense quantity of water.

Then it became clear that there was a large ice-dammed lake at Missoula in Montana. Old lake levels can be seen clearly – together with deer – on the hillside behind the house of our friends George and Barbara Stanley in Missoula. This Lake Missoula, as it is called, was up to 1100 feet deep (335 m) and covered an area of about 3000 square miles (7800 sq. km). The ice dam was at Spokane, near the Washington/Idaho border, at the mouth of the Clark Fork River. Like all good theories, the story has become more and more complicated, with arguments about how many times the dam may have given way. Successive gravels have different pebbles and flood debris covers an area of about 500 square miles (1300 sq. km) and reaches a thickness of at least 125 feet (nearly 40 m). Various gravels can be distinguished on the basis of the pebbles they contain, notably in the thick deposits of flood gravel to be seen in the canyon of the Snake River. These are overlain or interdigitated in places with loess, composed of dust blown across the dry Pleistocene plains. This is well seen in the Willow Creek section near La Crosse in Washington, but is really a matter for the next chapter. Modern thought suggests that it was probably not just a breaking of the ice dam, but *Jökulhlaups* or water from beneath the dam when hydrostatic pressure was sufficiently high (Waitt 1985 & Webster, personal communication 1987). It is estimated that this would have happened when the level of the lake rose to a critical level of about 600 m in depth.

Similar catastrophic flooding has now been recorded in many parts of the world. In a symposium on this phenomenon (Meyer & Nash 1987) examples were reported from the Great Lakes region, the Mississippi valley, northern Arizona and southern Utah, Virginia, West Virginia, Alaska, Peru, Iran, the Jordan Valley and Sinai. They were also supposedly recognized on other planets. It is always thus. When a new previously unknown phenomenon is recognized in one place, then people see it everywhere. This also happened, for example, with glaciations, turbidity currents, storm deposits and even extra-terrestrial impact craters, all of them 'catastrophic' in nature.

It is difficult to imagine the force needed to erode deep canyons in a rock as hard as basalt. Stephen Jay Gould has discussed the matter in his usual penetrating and witty style (1980, Chapter 19). He compared loosening sand grains by rubbing one's hand along the wall of a canyon (the gradualistic approach) with the sudden fall of a great block of sandstone from the famous Skyline Arch in Colorado. I have commented previously (Ager 1981a, p. 52) that such rare rock-falls and landslips may account for a major part of the wearing down of new mountain chains, rather than the traditional gradual processes of weathering. We have just heard (December 1991) of the tremendous rock-fall from one side of Mount Cook, New Zealand's highest mountain.

I remember my old departmental head (Prof. H. H. Read) commenting that he could lean against the Bank of England for the rest of his life and for centuries thereafter, but it would never fall down (see also Read 1949, p. 102). A more violent major natural force would be needed to destroy this heart of our economy.

Below the Missoula or Spokane flood gravels are the Bonneville deposits, which are 14 000 to 15 000 years old. These are thought to be the result of the breaking of a natural dam and the overflow of a great lake of which the Great Salt Lake in Utah is the last remnant. These waters flooded down the Snake River and spread beyond its banks, leaving extensive high bars of sand, gravel and boulders (Stearns 1962, Jarrett & Malde 1987). At least 17 floods are now postulated from this source together with eight separate episodes of catastrophic flooding from Lake Missoula, which Bretz suggested in his later work (1969), in place of the one gigantic flood which he postulated at first. Some geologists now claim as many as 100 floods altogether. So catastrophism in this area has become almost uniformitarian.

Similar rapid erosion of hard basalt occurred in the Grand Canyon of Arizona, though in this case by the action of a single river. One sight I would love to have seen was the spectacular 'waterfalls' of molten basalt which cascaded over the Esplanade Cliffs, forming four successive lava dams across the Colorado River. There may have been five more, of which only remnants are preserved. The river breached these dams remarkably quickly, cutting an average distance of 210 m back to the canyon wall. The cumulative distance of slope retreat was 1470 m – nearly a mile – per million years (Hamblin 1990, pp. 432–3). This retreat was intermittent and was controlled by tectonic uplift. Hamblin continued: 'erosion of the Grand Canyon did not take place at an imperceptibly slow and constant rate. . . . Erosion back to equilibrium occurs in a series of small pulses separated by relatively long periods of quiescence'. This is the principle I will emphasize over and over again in this book.

It is said of the Colorado River at the bottom of the Grand Canyon that it is 'too thick to drink but too thin to plough'. It certainly looked that way when I laboriously climbed down to it in 1959 (Figure 2.4) and then even more laboriously climbed up again. The sediment being carried by the river is rapidly filling up Lake Mead, behind what is sometimes called Boulder and sometimes Hoover Dam (dependent on whether the Democrats or Republicans are in office). Similar catastrophic flooding and rapid erosion of gorges probably happened in many other parts of the world (Meyer & Nash 1987), for example in the Kimberley Mountains of north-west Australia. We know only too well of the catastrophic effect of the breaking of man-made dams either due to human violence, as in the British bombing of the Ruhr dams in the Second World War or to human negligence. I think particularly of the failure of a dam in Bulgaria which drowned my friend Juli Stephanov (Ager 1972, p. 250).

Deserts

Mention of the early Americans probably drowned in the Missoula flood reminds me of one of the paradoxes that always please my black sense of humour. More people are drowned in the Sahara than die of thirst.

They are drowned by sudden torrents of water down the steep-sided channels or *wadis* (or *oueds* in former French North Africa). I know of two geologists who nearly did perish in this way, when their Land Rover stalled when crossing a wadi and a flash flood came down with great force. They had to climb on the roof of their vehicle and wait for the flood to subside. This happens because of the rare and short-lived rainfall in such deserts. My son spent three years in Burkina Faso, west Africa, on the southern edge of the Sahel, planning and supervising the building of small dams for scattered villages, to hold back the short-lived rainfall from flowing away eventually to the sea.

I was disillusioned when I had my first experience of deserts in the American south-west and later in North Africa, from Libya to Morocco. I suppose I was expecting sand-dunes and *Beau Geste* type forts. In fact most deserts, in my experience, are not sandy but have stony surfaces with the stones often shaped by the wind, (like the dreikanter on the book-shelf to my left) and such remarkable features as Mushroom Jebel (Figure 3.4) in eastern Libya. My most uncomfortable desert experience in that country was not getting stuck in the sand (though we did that as well), but painfully bumping over the ridges of gypsum that came up through the desert surface. In the Painted Desert of Arizona, I had difficulty in photographing my daughter (many years ago) standing on a typical desert surface that was not littered with Coca Cola cans (Figure 3.5). Waste disposal in hot deserts is always a problem, as it is in the

Figure 3.4 'Mushroom Jebel' in eastern Libya, cut by wind erosion in Miocene sediments, with my friend Harry Doust as scale. Photo DVA.

Figure 3.5   Typical desert surface in northern Arizona, USA, with my daughter Kitty as scale (many years ago). Photo DVA.

cold deserts of Antarctica. When I stayed in an oil company seismic camp in Libya, the air-conditioned trailers were surrounded by scree fans of beer cans (this was before the days of Colonel Khadafi and Muslim fundamentalism, I don't know what they drink now). Farther north on the coast, where the desert war swung to and fro for three years in the early 1940s, there are still heaps of rusty four gallon petrol cans, left behind by the British army (the Germans had the much more sensible and less perishable containers which we called 'jerricans'). There is no deposition over most of these deserts, so nothing is buried. Instead there is a hard crust, which is broken through by heavy weights and sometimes drapes over the side of escarpments (Figure 3.6). At the foot of Jebel el Gehenna (the 'hill of lost souls') in Libya, I took a symbolic photograph of a drill bit, left behind by oil men, just lying unburied on the desert surface (Ager 1981a, Plate 6.1). From the top of that jebel I was fascinated by the regular rectangles that extended across the desert surface like the marks left by a giant on stilts. These I found were left by heavy weights dropped from the back of trucks as a simple and cheap way to provide seismic waves.

   Vehicles too leave their tracks everywhere in deserts and I heard the story of a patrol in the North African campaign of the Second World War investigating some unusual tracks, only to find that they led to a

Figure 3.6   Desert crust draped over scarp of the Marada Formation (Miocene) in eastern Libya, with map case as scale. Photo DVA.

camp-site of the First World War identified by a crate labelled 1915. I have seen a very much older camp-site left by Palaeolithic people and recognizable by the ostrich egg-shells from the south, which they had used to carry water in this harsh environment. They had not been buried after all this time. At one place, some oil company joker had added to the effect by building a 'camp fire' of logs of Miocene wood, which also lie about on the desert surface; my brother brought some back from his wartime campaigning in both Egypt and Libya, and I later found plenty myself in the latter. A piece stands in a shallow flower-pot on one of my overloaded bookshelves above my left ear. Not far beyond Zagora in Morocco, the Sahara begins in earnest with the dunes of Tinfou. But, in spite of my preconceived ideas, sand-dunes are only very local in deserts, such as the isolated barchan dune in Libya, illustrated here (Figure 3.7). There are the 'sand seas' of Libya and Egypt, but they are still very much a minor part of the desert as a whole. One sees small shrubs or trees being buried on the depositional side of dunes and dead grizzled ones emerging, presumably years later, from the erosional side. The Grandes Dunes du Pilat near Arcachon in south-west France, said to be the highest in Europe (Figure 3.8) are busy burying a whole forest of living trees as they advance from the beach. One sees therefore, that blown sand is constantly on the move, encroaching, for example, on the

Figure 3.7   Typical isolated barchan sand dune near Zelten, eastern Libya. Photo John Shelton.

Figure 3.8   The 'Grandes Dunes du Pilat', south of the Bay of Arcachon, Gironde, France, moving inland and smothering a forest. Photo Alain Perceval. By kind permission of Editions d'art Yvon, Paris.

Roman city of Sabratha, near the Libyan coast. Nothing is permanent.

It is the same in North America where, in the dreaded Death Valley of California, dunes are restricted to a small area at the north end of this great depression between the mountain ranges. A grave near the main road through Death Valley, is of one Val Nolan; it records that he died 'a victim of the elements' in August 1931, but was not buried (and then presumably by human hands) until November.

One must come to the conclusion therefore, that desert sediments in the form of sand-dunes are continuously on the move and do not usually accumulate for the geologists of the future. What is more, desert animals and plants are unlikely to be preserved. This is in spite of the fact that there is more life in deserts than is usually supposed. I have seen gazelle tracks way out in the Sahara, although there is no obvious food for them for a considerable distance. It is well known that when there is a sudden heavy downpour, perhaps once in 10 years, then the whole desert surface bursts into fleeting flowers, as seen in that remarkable Disney film *The Living Desert*. A television film showed that a lizard was able to endure the heat of the desert surface (I think in the Kalahari) by raising one foot at a time for a little relief. I remember a student I had working in southern Morocco, who in desperation, sheltered each arm in turn in the shade of a very small shrub.

It is not generally appreciated, however, that deserts and tropical areas generally can be cold as well as hot. I have seen frost on the Indian Deccan and, when we first woke in the Painted Desert of Arizona, there was snow on the ground. My brother had some cold nights by his tank in the North African Desert in the Second World War as did my son in Burkina Faso in West Africa. A sergeant in the British Special Air Service died of exposure in Iraq during the Gulf War of 1991.

Therefore, though the desert may be uncomfortable, it is not impossible for animal life, including our own species, as exemplified by the Tuaregs and the Bedouin who, in oil-wealthy Saudi Arabia, now carry their camels from oasis to oasis in trucks.

Rivers

The most important terrestrial deposits at the present day are probably those of rivers. When a river or stream meanders, characteristically one side is being eroded, whilst on the other side sediment builds up on the convex bends. The term meander comes from the River Menderes in western Turkey and from the Greek word μαίανδρος to wander. Examples are common all over the world, one of the best being on the Tuolumne River in Yosemite National Park in California and the one I photographed from the air between Regina and Saskatoon in Canada (Figure 3.9). The actual channel of the river may be cut down into the underlying rocks or sediment and may contain torrential deposits of very

Figure 3.9   Meandering river seen from the air over Saskatchewan, Canada. Photo DVA.

coarse grain. Good examples are seen as one climbs up on to the Colorado Plateau near the Little Colorado River in Arizona. Here channels have been cut down into the soft underlying Triassic Moenkopi Formation (like the 'Keuper' of northern Europe) and the coarse torrential deposits filling the channel have led to their weathering out as ridges (Figure 3.10). Flattened pebbles commonly accumulate in an imbricate pattern with each pebble resting on the next in a downstream direction. The best example I have seen of this is along the Haast River in the South Island of New Zealand.

Inversion of the topography, as in Figure 3.10 is also seen in the great deposits of alluvial gravels which form cliffs, for example at Embrun in the French Alps, but these probably originated in the much harsher climatic of the recent past. Much depends, of course, on the angle of the slope. The extreme form are alluvial or scree fans, such as that behind Pralognan in the French Alps (Figure 3.11.). Scree, of course, represents the recent weathering of mountains, such as that below the Crêt de Bergers, near Veynes in the Alps (Figure 3.12). Such scree is particularly difficult to cross on foot. Its unconsolidated nature indicates that it formed very quickly and recently. On the other hand, it is noteworthy that in my beloved Jura Mountains (where I worked for a number of field seasons) the scree is usually comparatively old and is covered by a more comfortable soil for walking.

Figure 3.10   One of many torrential deposits filling channels cut in Moenkopi Formation (Triassic) along road up to Colorado Plateau, near Little Colorado River, northern Arizona, USA. Photo DVA.

Figure 3.11   Alluvial fan, Pralognan, French Alps. Photo DVA.

Figure 3.12   Fresh alpine scree, Crêt de Bergers, near Veynes, French Alps. Photo DVA.

Along the sides of the rivers are levées, where sediment has been piled up close to the river. These are seen particularly well along the sides of the Mississippi and along the River Loire in France, where one of the villages on the north side is called St Clément-des-Levées. I always think of levées when I see the Chinese/Japanese pictograph for a river, which is simply three parallel lines, one main one for the river itself and two smaller ones for the levées on either sides.

Rivers and levées, however, like everything else in this book, are ephemeral things. They are constantly on the move laterally destroying their own deposits. The outstanding example is the Hwang Ho or Yellow River in north China, which in less than 80 years moved its mouth some 250 miles (c. 400 km) from the Yellow Sea to the Gulf of Pohai on the other side of the Shantung Peninsula (see Ager 1981a, p. 53). It has been called 'China's Sorrow' because of its disastrous flooding with much loss of life and property. At the mouths of large rivers there are commonly accumulations of trees and other transported material washed down from upstream. The Mississippi provides good examples of this and such accumulations are well known in the geological record (see Chapter 4).

The most extensive river deposits are those of the flood plain, where the river has burst its banks and briefly spread over a large area, laying down silts and muds. If one drives across the delta of the Danube in Romania one is constantly aware of the dust. High plumes of it arise from every moving vehicle and it covers the windscreens (windshields) and side windows so that it is difficult to see. When I was there my party was constantly reassured that it was only so many kilometres to the tarmac (black top as the Americans call it) when conditions would be more bearable. This reminds one of the loess of the Late Pleistocene which is so thick along major rivers and will be discussed in the next chapter. I also think of the great 'dust-bowls' of Oklahoma and elsewhere in the USA during the 1930s when human interference made the soil surface unstable.

Also on deltas one may see the modern counterparts of 'coal measure' conditions. The best I have seen was on the Mississippi delta in Louisiana at Weeks Island, which is not an island but a gloomy swamp with swamp cypress trees *Taxodium distichum* trailing so-called 'Spanish moss' growing out of the murky water (Ager 1963, p. 164). Unfortunately it was too dark to photograph.

Lakes

Lakes are very varied in size and form. They are important receptacles for modern terrestrial sediments and life. Many of the smaller lakes around the world are markedly ephemeral. Even some of the larger lakes in hot climates vary considerably in size from time to time. Obvious examples are Lake Chad in north-central Africa and Lake Eyre in South Australia. There was once a number of elephants on a remote island in Lake Chad. This is clear evidence of how the lake has varied in size. It must have dried out sufficiently in Recent times to enable the elephants to reach the island. A remarkable pair of pictures form the frontispiece of *Phanerozoic Earth History of Australia*, edited by my friend John Veevers (1984). These show the Innamincka depocentre in south-west Queensland in its normal dry state and in its rare wet state after a flood.

Salts of various kinds, as in the Great Salt Lake of Utah and the Dead Sea, are an important feature of modern lake deposits, especially in desert areas. Sodium chloride has always been very important to man as a preservative and continues to be so in those parts of the world not yet reached by the refrigerator. When I was working in the western High Atlas of Morocco, salt would arrive in a lorry for the weekly market or *souk*. It was sold in screws of newspaper and was still red from its Triassic source, but that belongs in the next chapter. Not far away salt can still be seen forming along the sides of ephemeral streams in *oueds*. One sees it thus alongside the rough road that leaves the main road just north of the pass of Tizi n Tichka over the High Atlas. This minor road leads to the crumbling palace of the mighty Glaoui family at Telouet. These were the 'Lords of the Atlas', who ruled like mediaeval European monarchs, 'progressing' around their kingdom with their retinues, as enforced guests of their wealthier subjects. They used to have their enemies' heads on spikes round the battlements of Telouet. The last of this charming family attended the coronation of Queen Elizabeth II, as guest of Sir Winston Churchill, who convalesced at Marrakech, down on the plain below. The strength of the Glaoui family lay entirely in its possession of the vital salt, since Telouet stands on a major deposit and this was essential for preserving meat in the Moroccan climate. Meat in the little shops is likely to turn green overnight and, because of this, one of my students working there became 'hooked' on tins of sardines from Agadir and ate them for breakfast, lunch and dinner.

Far bigger deposits of salt occur in the countries to the south and great camel caravans still cross the Sahara to bring it back. They take 50 or 60 days to travel from Algeria to Bilma in Niger. A sign at the south end of Zagora, in the Anti-Atlas of Morocco, announces precisely 52 days by camel to Tombouctou (Timbuktu), though I would not advise anyone to attempt it because the frontiers today are probably closed. Timbuktu, though seemingly almost mythical, was a major centre of trade and culture in the past, largely because it stands at the head of a whole series of salt lakes. To the north, in Mali, salt is worked in opencast pits in the sterile plain of Tanezrouft. This was Lake Fagubine, the largest lake in west Africa, which has dried up in the last 10 years. Thousands of people were displaced as a result and a ship's anchor lies in the desert as a symbol of the past. Nothing could illustrate better the fleeting nature of desert lakes.

Smaller lakes with rapid evaporation also frequently deposit salts. A good example is Lake Magadi in Kenya where soda accumulates and is collected. In places the sodium carbonate forms stalactites, sometimes bent sideways by the wind. The soda encourages the growth of pink algae, which attract thousands of flamingoes to the lake. There are also modern diatomite deposits on the edge of the Rift Valley near Gil Gil. In Owens Valley, below the Sierra Nevada in California, there is a whole series of small lakes depositing a variety of different salts such as colemanite (hydrated calcium borate), attractive to my wife because she was a Coleman long before she was an Ager (Figure 3.13). Many modern lakes therefore are demonstrably ephemeral.

The modern salt lakes mentioned above presumably provide a key to the thick salt deposits of the past, but the mind boggles at the amount of evaporation they imply. Presumably they required repeated short-lived events, when a lake formed over and over again and was evaporated to dryness. Alternatively ordinary sodium chloride deposits may be explained by a sill over which a lake was repeatedly replenished by the sea. We see a modern example of such a situation in the Zaliv Kara Bogaz Gol or embayment on the east side of the Caspian, where salinities are exceptionally high. A similar situation is seen in Baffin Bay, an embay-

Figure 3.13   Dried up salt lakes, north end of Owens Valley, California, USA. Photo DVA.

ment off the Laguna Madre on the coast of Texas. The latter is almost cut off from the sea and is hypersaline, the former reaches salinities of more than 80 parts per 1000 (Ager 1963, p. 239). Again an explanation of short-lived and repeated episodic events seems unavoidable. However, I stray into the next chapter.

More 'normal' non-salt-depositing lakes have probably not been so thoroughly studied, though excellent work has been done in the Great Lakes of North America, notably by my former colleague Jack Hough (1958) of Illinois. Though many of them, such as the Great Lakes and those of East Africa may seem permanent to our eyes they are, like so many things in this book, ephemeral. As soon as they form, they start filling with sediment. The velocity of rivers and streams entering a lake is immediately reduced in the static water and so they tend to deposit their loads. I have already commented on how man-made Lake Mead, behind Boulder Dam is filling with sediment from the Colorado River. Natural dams which may hold up a lake are often rapidly eroded and give way as with the basaltic dams in the Grand Canyon mentioned above. Other natural dams are formed by great rock falls and mud flows. For example, a great landslip occured in 1442 in the French Sub-Alps at Claps-de-Luc near Die (pronounced 'dee', never say die!). It must have temporarily blocked the River Drôme, long after Hannibal passed that way.

A special case, though very common in volcanic areas, is the crater lake, filling the depression left by an eruption or series of eruptions. Crater Lake in Oregon is a particularly well-known example of one of these. European examples are the *Maars* of the Eifel district in Germany, such as the Lochersee, formed in a great crater of a peléan type eruption during the post-glacial Allerød oscillation. In the French Auvergne there are lakes such as the Lac d'Aydat south of Clermont-Ferrand and west of Naples in Italy is Lago di Averno, the classical entrance to the underworld. On the beautiful, but tourist infested island of Bali, the crater lake on Mount Batan is used for water skiing.

Some of these, such as the one in Oregon, are calderas rather than simple volcanos, as is the Ngorongoro Crater in Tanzania. The very steep walls of this collapsed caldera provide what is probably the best nature reserve in the world, with nearly all the famous East African fauna contained within its high walls. I did not really believe in the rhinoceros until I saw them here. Only the giraffe apparently found the walls too steep to descend. The lake here, however, is comparatively small, but provides a home for a number of hippopotami. 'How did they manage to get in?' asked a bright local student, but did not get a reply from the director of the trip. I can only presume that they rolled! In the North Island of New Zealand two large lakes, those of Rotorua and Taupo fill old collapsed calderas. Tufa is abundant around the latter and one of my party tricks for visitors is to throw a large boulder from there into our fish-pond. It floats! The most famous deposits of Rotorua, however, were the pink and white terraces which were the chief tourist attraction in New Zealand before they were destroyed by the eruption of Mount Tarawera on 10 June 1886. Also buried were two Maori villages, one of which was still being excavated as a southern equivalent of Pompeii when we were there exactly 100 years after the eruption (see Chapter 11).

Lake Nyos, a crater lake in Cameroon, with an unfortunate and unusual recent history, is discussed in Chapter 12. Another crater lake is seen at the southern end of the Satsuma Peninsula in Kyushu, Japan, near the perfect volcanic cone of Mount Kaimon (sometimes called 'Kyushu Fuji'). This lake is said to contain a monster named 'Issy', to match 'Nessy', the supposed monster of Loch Ness in Scotland. The latter lake, however, is of a quite different origin, having been gouged out by the ice in the Pleistocene. The terminal moraines produced by glaciation often dam up lakes. A splendid example of a moraine-dammed lake is Lake Wallowa in Oregon near the town of Joseph, (commemorating the Indian chief, 'Old Joseph', who practised peace and did not resist the white man with the sad result that his son 'Young Joseph' was forced to make an heroic march with his tribe some 1800 miles to sanctuary in Canada). A lateral moraine of this lake is illustrated (Figure 3.14). Another is beside the Wittenhorn Lake in western Germany. Well-

Figure 3.14   Lateral moraine, Lake Wallowa, Oregon, USA. Photo DVA.

known examples of moraine-dammed lakes much closer to my home are the small lakes around the mountains of Snowdon and Cader Idris in North Wales, such as Llyn-y-Gader on the west side of the highest mountain in Wales, i.e. Snowdon.

Ox-bow lakes, as is well known, are small arcuate lakes left behind by the constant changes in meandering rivers. A good example is shown in Figure 3.15 photographed over Arkansas. Good examples closer to home are seen along the River Tywi near the village of Bethlehem (that's our Welsh one, which is very popular for its postmark at Christmas time).

Glaciers

Glacially dammed lakes have already been discussed. Glaciers themselves are a comparatively minor element on the Earth as it is today, but the study of the deposits which they produce provided the vital evidence that led scientists such as Agassiz to postulate the much greater glaciations of the past. Thus the jumbled rock debris at the end of the Rhône glacier in Switzerland and those of the Mer-de-Glace and the other glaciers near Chamonix in France (Figure 3.16), enabled the interpretation of the tills of the past. In other words they were the evidence which led to the

Figure 3.15   Ox-bow lakes from the air over Arkansas, USA. Photo R. Balsley, USGS.

recognition of the catastrophes of the Pleistocene glaciations. Glaciers are found in every major mountain range. I have seen them in the Alps (notably the Rhône glacier), in Norway (notably the Briksdal glacier, said to be the largest in Europe, Figure 3.17), in New Zealand (notably the Franz Josef and Fox glaciers in the Southern Alps) and the Athabasca glacier (part of the Columbia icefield) in the Canadian Rockies. All are slowly moving onwards and, particularly in the last few years, melting at their snouts. An indication of their rate of movement was provided by the recent discovery of a Bronze Age man at the end of a glacier on the borders of Switzerland and Italy.

Hot springs

Hot springs are very local phenomena. I think of the one near Embrun in the French Alps which produces a gutter of hot water raised on its

Figure 3.16   Till below glacier; Chamonix in background, French Alps. Photo DVA.

Figure 3.17   Main Briksdal glacier, Norway. Photo DVA.

own deposits. Others such as those at Karlovy Vary in Czechoslovakia are the curse of local plumbers as they block up pipes with their deposits. They also petrify items such as the rose on another of my book-cases. When the town was called Karlsbad before the Second World War, it was a great resort for the aristocrats of Europe, who drank the supposedly beneficial waters. After the war it became a spa for worthy workers. When I was last there, Mrs Kruschev was staying, paradoxically, at the Imperial Hotel, overlooking the town. Her husband fell from power the next day. There are also sulphurous springs at places such as near Zell-am-See in Austria. Closer to home is the hot spring at Bath, which has been used for medicinal purposes and for bathing since Roman times. I have a private theory that it may be related to volcanicity (not seen here at the surface) but such as occurs at comparable rift margins in continental Europe.

We should also include here the geysers of volcanic areas such as the Rotorua area of New Zealand, Iceland and the well-known 'Old Faithful' in Yellowstone National Park, Wyoming. I am always unfortunate in missing the sudden spurting of geysers; perhaps I am too impatient. Siliceous deposits of geyserite are extensive in these areas. On the Tihany Peninsula, which projects into Lake Balaton in Hungary, there are said to have been more than 30 geyserite-producing springs around a supposed collapsed caldera.

Caves

Cave deposits are, of course, even more of a special case, largely (though not entirely) restricted to limestone country. The stalagmite and stalactite in caves hardly need discussion here, but there are other deposits such as the mammoth dung deposits of Bechan Cave, Montana and the thick layer of bat droppings in the Mammoth Cave of Kentucky. I was told proudly that they used this to make gunpowder in the American War of Independence. I was secretly amused at this as one of the foundations of American liberty!

4
Ancient terrestrial deposits

Ancient terrestrial deposits are chiefly the sediments laid down by meandering rivers and in ephemeral lakes and swamps. There are also some blown sand and dust accumulations which are probably more common than is generally supposed. The latter are, for the most part, accumulations in ancient deserts or on broad dry plains beside major rivers. There are also the deposits associated with glaciations. All of them were brief episodes in geological terms.

Deserts

One of the most startling and 'catastrophic' discoveries of recent years was that *The Mediterranean was a Desert*, which is the title of a very readable book by Ken Hsu (1983). In this he gives a blow by blow account of this remarkable discovery as it was confirmed on a voyage of the *Glomar Challenger* by many borings into the sea-floor. These proved considerable thicknesses of evaporites and other terrestrial deposits all dated as being formed at the end of Miocene times. Evaporites of this age are also seen in many of the lands bordering the Mediterranean. I have been impressed by them particularly in northern Libya and in Sicily. It seems that the Tethys Ocean, which played such an important role in the history of this region, dried out completely at the end of the Miocene (in the Messinian). Maria Cita has long drawn attention to the 'Messinian Salinity Crisis' which had a catastrophic effect on the marine faunas of the region until they all eventually disappeared. This was clearly demonstrated and summarized in her 1982 paper. It was also suggested that as a result of the fall in sea-level, there was downcutting in the Italian Alps, to produce the beautiful Italian lakes such as Como and Maggiore (see Bini *et al*. 1978, Finckh 1978, Cita *et al*. 1990, Hsu 1991). These were previously presumed to be glacial in origin. The same origin has been postulated for the Var and Durance 'canyons' in southern France (Clauzon 1978 & 1979). Earlier Miocene deposits were fully marine, as is seen in the Marada Formation of eastern Libya. That formation can be seen to become more marine as one traces it northwards towards the present Mediterranean, with an abundance of echinoids, bivalves, barnacles and other marine invertebrates. Southwards it passes into entirely continental deposits with root traces and silicified palm trees (Doust 1968, Selley 1966 & 1969).

In Chapter 3, I pointed out that organisms are more common in modern desert environments than might be expected. So far as the ancient record is concerned, such organisms are not often preserved. But they do occur. Thus the famous Permian Coconino Sandstone of the Grand Canyon, with its strong cross-bedding (Figure 4.1) is generally accepted as an aeolian deposit, but it does contain abundant vertebrate and invertebrate trace fossils. These appear to have been formed chiefly in dry sand, and include some tracks very like those of the modern scorpion (Middleton *et al.* 1990). They appear to have had the perverse habit of always walking up the sides of dunes but never down! This again is probably a matter of preservation (see Chapter 2) with the downhill tracks being obliterated by sliding sand on the depositional side of the dunes.

Figure 4.1   Cross-bedding (up to 20 m amplitude) in Coconino Sandstone (Lower Permian), Bright Angel Trail, Grand Canyon, northern Arizona, USA. Photo DVA.

My friend Hilda Stoneley (1958) described an extensive flora in the Permian Hilton Plant Beds of north-west England, which are clearly of desert origin. All the sedimentary evidence, both above and below the plant beds, is of such an environment, with red beds, cross-bedded sands and evaporites. In an earlier paper (1956) she had described a new plant genus *Hiltonia* in which the leaves had unusually thick cuticles. This feature characterizes plants growing today in hot, dry habitats, since they protect the plant from excessive loss of moisture.

In Europe the best known ancient terrestrial deposits are those of the 'Old Red Sandstone' in the Devonian and the 'New Red Sandstone' in the Permo-Triassic. It is a common mistake among non-geologists to equate the 'Old Red Sandstone'* with the 'New Red Sandstone'. The only thing they have in common is that adjective 'red'.

There is no doubt that much of the 'New Red Sandstone' of Europe represents desert conditions, with dune sands, loess-like deposits and evaporites, but the 'Old Red Sandstone' has none of these. It consists rather of flood plain deposits with land plants and ephemeral lake deposits with fish and the branchiopod crustacean *Cyzicus* (formerly called *Estheria*) which still lives in short-lived pools at the present day.

The land plants of the Devonian were known primarily (at least to me) by the remarkable silicified flora of the Rhynie Chert in north-east Scotland. These rocks were originally thought to be volcanic, since there are many volcanic rocks in the neighbourhood, but when sectioned they revealed excellent cross-sections of well-preserved plant tissue. These plants evidently grew in a marsh or bog and the silicifying fluid emanated from the local volcanics. They are some of the earliest land plants known and still probably needed water to support them. Since then many land plants have been found in Silurian and Devonian deposits, notably those very early land plants found and studied by my friend Dianne Edwards in the 'Old Red Sandstone' of South Wales (see Edwards *et al.*, 1992).

It should be pointed out that continental deposits of Devonian age extend all the way from the British Isles via Czechoslovakia to the former Soviet Union (where it is still known as Old Red Sandstone). There are also continental deposits of this age in eastern North America including, in this case, evaporites which continue the story of the great evaporite basin around the Great Lakes in Silurian times. There were also forests as in the Gilboa Formation of the Appalachians. Similar beds in Greenland have yielded the remains of the first known amphibians to have crawled from the sea.

The 'New Red Sandstone' is often neither red nor sandstone and it is certainly not new. That last adjective is commonly applied to the second oldest phenomenon of a particular type, be it a college, an inn or a river! The Permian part of it is particularly well-known in Europe for its valuable and varied salt deposits, notably those at Stassfurt in Germany, where evaporation must have reached the ultimate stage with the deposition of various potash salts. Diapirs formed by rising salt also provide valuable oil and gas traps.

---

* The ORS has always been the 'Old Red Sandstone' to geologists, but now it also means 'Overseas Research Student' or 'Oral Rehydration Salts' (to help desiccated refugees). In my college it now also means 'Observational Research Swansea'. So much for 'apt acronyms artful aid'!

Continental Triassic beds are even more extensive in Europe and North Africa. They continue the same story with much salt and gypsum. This is seen, for example, at Jebel Amsitten in the western High Atlas of Morocco (Figure 4.2) where the salt has only just emerged from the adjacent red rocks (unfortunately coloured illustrations are too expensive!). At the bottom of the Triassic, at least from Britain to Bulgaria, there are commonly coarse breccias or conglomerates. At Ogmore-by-Sea, near my home in South Wales, there are obvious wadies filled with short-lived torrential deposits coming off a nearby island of Carboniferous rocks like those on the Colorado Plateau I mentioned in the last chapter (see Figure 3.10.) The Buntsandstein is more extensive and is widely seen as a building stone, for example at Koblenz in Germany, where the Moselle joins the Rhine. The 'Keuper' or Upper Triassic of northern Europe is mainly a loess-like dust deposit, blown across the dry land surface before grass had evolved to stabilize the soil. The European Keuper is mirrored in the Newark 'Super-Group' of the eastern sea-board of the USA, in a series of small faulted and short-lived basins. The main difference is that continentality in the American basins went a stage further with the development of coals. There are also volcanic extrusions and intrusions representing the first cracking open of the Atlantic Ocean, though these are now known to range into the Jurassic. Similar igneous rocks are only seen in the south of Europe, notably in Spain and Portugal. In Morocco (Figure 4.3) they conventionally mark the top of the

Figure 4.2   Salt pans derived from the adjacent Triassic, Jebel Amsitten, western High Atlas, Morocco. Photo DVA.

Figure 4.3   Basalt flow near top of Triassic, Agoum near Ouzarzate, southern side of the western High Atlas, Morocco. Photo DVA.

Triassic. One always knows when one gets on to the Triassic there because of the small boys at the road-side selling amethyst from the basalts, sometimes 'improved' with day-glo paints.

It is always difficult to correlate the deposits of terrestrial basins with those of marine sequences. Palynology has helped a lot in recent years but has its limitations. I have suggested, in connection with the European Triassic (Ager 1970) that marine transgressions in one part of a continent must have a widespread effect in the non-marine areas, especially in the centre of depositional basins. Thus the important transgression of the Late Muschelkalk (Ladinian) on the European continent is, I think, reflected in the 'Waterstones' of the English terrestrial Upper Triassic. This was 'the most important event in the Triassic history of Europe' (Ager 1981b), but – as so often in my story – it was a very brief one.

The Triassic Moenkopi Formation of Arizona is almost incredibly like the European Keuper and similar red beds of this age are seen in countries as far apart as England, Germany, as around the Black Forest and Harz Mountains (Figure 4.4), Bulgaria, Argentina and I have seen something very like it in China. In its upper part the Keuper 'Marl', or Mercia Mudstone Group as we must now remember to call it in Britain, commonly shows alternations of red and green beds with layers of gypsum. It is presumed that the two colours represent alternating oxidizing and reducing conditions. Again nothing lasted long. These alterna-

Figure 4.4   Keuper (Upper Triassic) mudstones and thin sandstones, comparable to the Moenkopi Formation of Arizona, north of Polle in the Harz Mountains of Germany with my son Martin as scale. Photo DVA.

tions culminate in a wholly green member called the 'Tea Green Marls' in Britain and the 'Marnes lie de vin' (wine dreg marls) in France, reflecting no doubt the different drinking habits of the two countries. These were the final terrestrial deposits before the sudden invasion of the Rhaetian sea at the end of Triassic times.

Swamps

I should before now have mentioned the Upper Carboniferous (West-phalian in Europe, Pennsylvanian in North America), but it was logical to discuss the Devonian and Permo-Triassic (before and after the Car-boniferous) together. The extensive coal and associated deposits of the Upper Carboniferous are almost incredibly widespread, from the Ameri-can Mid-West to the Donetz Basin in what was (until recently) the Soviet Union. This is one of those examples of persistence of facies on which I have often commented before. The coal measure swamps were formed on the tops of deltas with a dense development of forests and smaller plants. In reconstructions of the coal measure environment, there always seem to be one of those rare early amphibians peeping between the trees. In fact there is quite an abundant fauna to be found, for example at the famous locality of Mazon Creek in Illinois. Jarzembowski

(1987 and later) described remarkably well-preserved insects (including some with their colours still visible) from the Upper Carboniferous Coal Measures of Writhlington in the west of England and Todd (1991) described a whole community of forest litter animals from the same locality. This is another of those near miraculous preservations, such as the Mid Cambrian Burgess Shale of the Canadian Rockies (which obstinately refused to yield me anything when I was there) and the Solnhofen Lithographic Stone of south-west Germany, which give us a fleeting glimpse of what life was really like in the past.

The coal measures are well-known for their cyclothems or cycles of sedimentation with rapidly alternating conditions from marine beds through deltaic sandstones to coals and beds of freshwater bivalves. From Swansea we used to take our first year students on a traverse of the classic South Wales coalfield early in their course; but it was very difficult to show them coal itself on the surface. This indicates the short-lived nature of coal formation. Coal seams, like so many other things in earth history, were formed in very brief moments, geologically speaking, even the thickest of them. One wonders how many generations of forest were involved in the formation of thick coal seams, such as the 'Thick Coal' in South Staffordshire, England, between 6 and 10 metres in thickness, which led to the industrial development of Birmingham and that region. Farther north, the coal splits with several thinner coals in place of one thick one and these equivalent strata are about 50 metres thick. So a lot may be packed into a very little. In the famous mining valley of the Rhondda in South Wales (now sadly without coal mines) there are the No. 1 Rhondda and No. 2 Rhondda seams. Below these are five more seams totalling some 9.3 metres of coal altogether. The forests must have grown repeatedly but briefly on the deltaic swamps of the Westphalian. Similar coals occur in the Permian of Gondwana, notably the Ecca Series in South Africa and in Queensland, Australia, I hear of seams in the latter up to 100 feet (c. 33 m) thick and readily accessible from the surface. It is obvious why our deep though good coal in Britain cannot compete in a free market.

Probably the most convincing proof of the local rapidity of terrestrial sedimentation is provided by the presence in the coal measures of trees still in position of life. Two Late Carboniferous trees stand in the garden of Swansea Museum. They were found at Nant Llech in a tributary of the River Tawe which gives Swansea its Welsh name (Abertawe or mouth of the Tawe). An original drawing of them in position is shown in Figure 4.5. They were found by William Logan, later Sir William Logan and Director of the Geological Survey of Canada. Such standing trees are not uncommon in the Upper Carboniferous. A photograph of another standing tree – a Sigillaria – from Abertillery in South Wales, is illustrated in Bassett and Edwards (1982) and in Glasgow there is a famous 'fossil forest' in Victoria Park, where Late Carboniferous stumps with

Fossil Segillaria, discovered in Cwm Llech Vale of Swansea by M: Logan.

Figure 4.5   Old print showing fossil trees in position of growth at Nant Llech in the Swansea Valley, South Wales. The trees are now preserved outside Swansea Museum.

their main radiating roots were discovered during the quarrying of an overlying dolerite sill.

Broadhurst and Loring (1970 and earlier papers by Fred Broadhurst) recorded standing trees up to 10 m high in the Lancashire coalfield of north-west England. They showed that rapid sedimentation had alternated with slow, rather like ballroom dancing – 'slow, slow, quick quick, slow'. Obviously sedimentation had to be very rapid to bury a tree in a standing position before it rotted and fell down. David Smith of BP did an instant calculation when I had talked about these things, as to what this meant in terms of rates of sedimentation (personal communication 1988). I later did my own calculation and it proved even more surprising. If one estimates the total thickness of the British Coal Measures as about 1000 m, laid down in about 10 million years, then, assuming a constant rate of sedimentation, it would have taken 100 000 years to bury a tree 10 m high, which is ridiculous. Alternatively if a 10 m tree were buried in 10 years, that would mean 1000 km in a million years or 10 000 km in 10 million years (i.e. the duration of the Coal Measures). This is equally ridiculous and we cannot escape the conclusion that sedimentation was at times very rapid indeed and that at other times there were long breaks in sedimentation, though it looks both uniform and continuous.

Another Carboniferous tree, though older than the ones above, stands in the grounds of the Natural History Museum in London. In fact it stood first for hundreds of millions of years in the Scottish Coal Measures before it was transported to London. It only fell when it was knocked down by a bomb during the Second World War. After the war it was restored to its previous upright position, but only with the help of a lot of cement.

Broadhurst, Simpson & Hardy (1980) also demonstrated episodic sedimentation in the English Upper Carboniferous (Westphalian) by means of the non-marine bivalves, which are often common in these deposits. They plotted the number of 'escape shafts' of these molluscs, when they had to burrow upwards to avoid fatal burial. Such escape burrows occur only in sandstone layers, which are interbedded with siltstones. The authors concluded that these sandstones were deposited very rapidly in wet monsoon seasons, whilst the siltstones accumulated slowly during the dry seasons.

Standing trees are known at many levels and in many parts of the world. Thus there are trees in the Devonian of Gilboa in New York State, mentioned above, which were evidently buried by a sudden rush of sediment. In the English Upper Jurassic we have the 'fossil forest' on the Dorset coast at Lulworth Cove, where cycad-like trees stand in an ancient soil and are encased in tufa. Unfortunately the locality is on a tank firing range (where I did my share of shooting in the early 1940s) and it is not easy to visit. At Yellowstone National Park in Wyoming, a whole forest of Miocene trees is buried in volcanic ash. Similarly in the *Ginkgo* Petrified Forest near Vantage in Washington State, many different Miocene trees are still in position, preserved between lava flows.

Rivers

The most obvious evidence of ancient rivers is provided by channels cut down into the underlying sediments. The colloquial term 'wash-out' comes from the coal-miners' expression for when a coal seam is washed out by such a channel. My former colleague Gilbert Kelling was able to work out the whole drainage pattern of the South Wales coalfield at one moment in Late Carboniferous times on this basis (Kelling 1968).

Late Eocene in age were the London Clay deposits of the Isle of Sheppey in the Thames estuary, which appear to have accumulated at the mouth of a large tropical river, comparable to those seen today in south-east Asia. These beds have an abundance of a very diverse fauna and flora, ranging from the logs, twigs and fruits of land plants (such as the palm *Nipa*) to marine gastropods and fish (such as the palate teeth of a ray *Myliobatis toliapicus*, which I use as a paper weight). The concentration of fossils here does not necessarily imply a long period of time, when one thinks of the super-abundance and diversity of life in the tropics. It is noteworthy that the logs are commonly riddled with the borings of the 'ship-worm' *Teredo* (actually a bivalve mollusc). Another former colleague, Dick Selley (1969) described and figured a meandering estuarine channel in the Miocene of Libya.

As mentioned in the previous chapter, there are often great accumulations of logs at the mouths of great rivers such as the Mississippi. These have obviously floated down the river before the flow of the water was retarded. They cannot have been there long, however, or they would have rotted away or been riddled with borings as are those at Sheppey, mentioned above. A comparable ancient example is provided by the so-called 'pine raft' of Early Cretaceous age at Hanover Point in the Isle of Wight. A more famous example is the accumulation of Late Triassic logs in the 'Petrified Forest' of Arizona, which have been transformed into beautifully coloured agate ('fossilized rainbows' they are called locally). In both these last cases the wood is mainly araucarian ('monkey puzzle' family) and only battered logs are preserved (Figure 4.6) with no twigs or leaves, which were presumably lost in the rough and tumble of transport from the upper reaches of the rivers concerned.

McKnight *et al.* (1990) described an Upper Jurassic fluvial deposit in the Junggar Basin of Xinjiang, China (Tibet) with silicified logs at two levels and stumps in life position at another. Evidently these record periodic river floods. Similarly Hunt (1991) described the effects of Late Cretaceous large scale flooding of a forest in New Mexico. Rotted tree stumps are common in flood plain deposits; logs are more common in

Figure 4.6   Battered, river-transported Triassic log, 'Petrified Forest', northern
Arizona, USA, with my daughter Kitty as scale (again a long time ago!). Photo DVA.

channel sandstones, mainly oriented normal to the direction of current
flow. Many disarticulated vertebrate bones were found, most closely
associated in the river channels. Dinosaur bones, however, are more
common in the overbank deposits but include what the authors called a
'dinosaur graveyard', which appears to be the infilling of a channel.
Evidently there were several earlier floods, but the one finally responsible
was a short-term, high energy affair.

Returning to the 'Old Red Sandstone', it was mentioned previously
that much of this was probably deposited on the flood plains of rivers.
Similarly much of the European topmost Keuper deposits in Late
Triassic times probably represent fine wind-blown dust on flood plains.
Perhaps the best known and most extensive flood plain deposits are those
of the loess, which is now thought to have been laid down very quickly
in Late Pleistocene times. An early European visitor to China compli-
mented the locals on their industry in removing all the stones from their
soil. The Chinese certainly are industrious, as can be seen from the large
numbers of men and women tilling the fields with hand-held hoes and
excavating new roads by hand, but they cannot claim to be that industrious. The visitor was looking, of course, at the vast areas of fine-grained
loess which extend away from the great rivers of that country. One
imagines icy Pleistocene winds blowing across dry flood-plains.

I wrote in the last chapter of the dusty Danube delta. The Danube
valley was evidently equally dusty back in Pleistocene times, as can be

seen from the high cliffs of loess along that river in Romania. One of the most remarkable geological sights I have ever seen was at Mikulov in Czechoslovakia where an excavation in Danubian loess shows the remains of literally dozens of mammoths (Figure 4.7). I was puzzled then (and still am) whether this was one of those mythical elephants' grave-yards, where the great beasts went to die, or whether it was a herd of mammoths suddenly buried in blowing dust from the river flood plain. My catastrophist spirit inclines to the latter. The other great rivers of Europe also have their extensive loess, notably the Rhine and the Rhône. Even the Thames has a little loess at Pegwell Bay, first recognized by my former colleagues Pitcher, Shearman & Pugh (1954). I was puzzled by a small patch of loess near Wellington, New Zealand, since there are no large rivers there, even of Thames proportions. Sir Charles Fleming, however, suggested that it came from a broad coastal plain when the sea-level was low during the Pleistocene (personal communication 1986).

Lakes

Another kind of terrestrial sedimentation is seen in the Middle Eocene Green River Formation of Wyoming, Colorado and Utah. This consists of lake deposits, often distinctly varved. They are famous for their well-preserved fossil fish. They are about 2000 feet (c. 610 m) thick and cover an area of about 25 000 square miles (65 000 sq. km). The varves suggest that the climate alternated between long warm summers and cool moist winters (Bradley 1948). The varves in various beds vary from a minimum of 0.014 mm in rich oil shales (when organic material accumulated slowly) to about 9.8 mm in fine-grained sandstones (which must have accumulated much more rapidly). They were deposited in a thermally stratified lake which may not have been more than 75 to 100 feet (c. 30.5 m) deep when the varves were formed and the differences in thicknesses may have been accentuated by differential settling rates. Assuming that the varves were annual phenomena, then the deposition of the Green River Formation may have lasted five to eight million years. It is preceded and followed by fluviatile deposits. Deposition therefore was comparatively rapid. Lake deposits in the Scottish Old Red Sandstone (discussed above) are famous for their concentrations of fish, presumably killed in some local catastrophe such as the sudden drying out of the lake in which they lived.

One of the most fascinating areas I know as a palaeontologist is the Pannonian Basin of Hungary and neighbouring countries. This was first an arm of the sea in Miocene times, but was then broken up into a series of lakes. The changing salinities produced what I have called an evolutionary hot-house, because the molluscs present in the lakes changed so quickly. The classic example is the smooth, freshwater snail

Figure 4.7   Excavation of Mammoth remains preserved in loess, Mikulov,
Czechoslovakia. Photo DVA.

Limnaea which changed quite suddenly into the limpet-like, strongly ornamented form *Valenciennesia* (Ager 1963, p. 279). Another strange mollusc here is *Congeria ungulacaprae* about which the local children tell a story about a goat-herd and the god of the lake.

The goat-herd loved the same woman as the god of the lake, who rose up and drowned the goat-herd and his flock, which are now recognized by the twisted shells of *Congeria*, interpreted as the goats' horns. The history of that lake was recently summarized by Kázmér (1990). Whilst these normally slowly evolving molluscs evolved very quickly, the normally rapidly evolving horse family was represented by a single species of *Hipparion* grazing placidly and unchanging round the borders of the lake. Nothing could demonstrate better the way a sudden environmental and geographical change can cause a rapid evolutionary change.

Salt lakes were discussed in Chapter 3, but nothing today compares with the immense thicknesses of evaporites in the past. Noteworthy is the great salt basin of Late Silurian age in the Great Lakes region of the USA. The amount of water that needed to be evaporated to produce this amount of salt is hardly imaginable. One must conclude that this was a lake basin that was repeatedly replenished by the sea, presumably over a sill of some kind. Basins such as that at Stassfurt in Germany contain potash as well as sodium salts, which implies a further stage in evaporation. Gasse & Fontes (1989) described a Quaternary salt lake in the Afar triangle of north-east Africa which showed evidence of exceptional floods with repeated marine infiltration.

Non-aqueous environments

Grasslands, such as those patronized by the horses, were a new feature of the terrestrial environment in Tertiary times, but they and their animals stand very little chance of preservation. I was struck in East Africa by the short-lived nature of animal remains on the grassy surfaces. Skeletons remained unburied, notably those of the gnu (wildebeest), buffalo and elephant, which soon crumble away (the elephants had already lost their tusks to the poachers). So we get a bias in the potential preservation of terrestrial environments, with very little to be seen of the grasslands, but plenty of the muddy lakes and swamps (see Chapter 2).

Forests too have a poor preservation potential unless they grew in swamps like the coal measure forests already discussed. Peats are another matter. They form everywhere from the shoreline to upland areas. Ancient black peats tend to alternate, for reasons not quite clear to me, with layers of white kaolin clay. I saw this in the Cretaceous of Minnesota (Figure 4.8) and later in strata of the same age in southern Saskatchewan. When I noted these contrasting layers I was immediately struck by their resemblance to the Mid Tertiary Bovey Tracey Beds of my homeland. These formed in a small basin near the Dartmoor granite of South

Figure 4.8   Cretaceous kaolin clays and lignites, Minnesota, USA. Photo DVA.

Devon, from which the kaolin was presumably derived. There are plenty of granites to hand in Minnesota too, in fact the beds lie directly on one, but I was puzzled in southern Saskatchewan, where there did not seem to be a granite outcropping within hundreds of kilometres. No doubt it was different in Cretaceous times.

Ancient peats and brown coals are found in many parts of the world and at many different stratigraphical levels. They are important economically in many countries. They all seem to represent brief episodes of bog-like conditions in which plant material accumulated. In Sweden they occur in the top-most Triassic and lowest Jurassic and are mined at Vallakra near Halsingors in what must be the cleanest coal-mine I have ever visited, as is characteristic of that clean and tidy country. In Scotland we have the Upper Jurassic Brora coal and the main coals in the west of the USA and Canada are Cretaceous in age. In Hungary too there are Early Jurassic coals in the south and Eocene ones as well, where they are seen at Bagolyheg in the Vertés Mountains, resting unconformably on Cretaceous bauxite (Figure 4.9).

Bauxites are a product of ancient sub-aerial weathering of aluminium-bearing rocks under tropical conditions. The name comes from Les Baux in Provence, which was the 'nest' of *Une race d'aiglons* (a race of young eagles), who were the most powerful feudal lords in the region during the Middle Ages and were followed by the brigand Raymond of Turenne who terrorized the neighbourhood. The strength of the site came from its topography, perched on a detached spur of the Alpilles range, rather than from its bauxite, which is worked for aluminium (Figure 4.10) and

Figure 4.9   Eocene brown coal resting unconformably on Cretaceous bauxite, Bagolyheg, Hungary. Photo DVA.

Figure 4.10   Bauxite at the 'type locality' of Les Baux, Provence, France. Photo DVA.

has replaced brigandry as the basis of local industry. Bauxites are seen in many parts of the world and provide good evidence of past, usually fleeting episodes of tropical sub-aerial weathering. In Minnesota, I was very impressed by a section near Redwood Falls which showed the Archaean Morton Gneiss of the Canadian Shield weathered to a depth of about 30 m below transgressive marine Cretaceous. This must have been the effect of sub-aerial tropical conditions in earlier Mesozoic times (Figure 4.11).

Kaolin or china-clay also represents the deep weathering of felspar-bearing rocks and is commonly found associated with granitic rocks, such as the Permian Dartmoor Granite in Cornwall. I knew an old sailing ship captain who used to take it from Looe all over the world for the making of porcelain and china. I have seen it too in the Cretaceous at Langley in South Carolina, where it was presumably derived from the ancient rocks of the Piedmont, and in the Miocene of the Pugu Hills in Tanzania, where it is worked in horizontal adits and in the section illustrated (Figure 4.12) includes a thin red band probably representing the brief extension of a flood plain. Here it was close to the granitic rocks of the African Shield.

Diatomites of freshwater origin are seen in the Cretaceous of Lebanon containing abundant perfectly preserved fish known to Alexander the Great and his army as he passed that way. I collected a surprising number of fine fossil fish in a remarkably short time from a Miocene diatomite in Sicily. In Romania similar fish are found with their mouths open as they died of suffocation in an Oligocene diatomite (Ager 1963, p. 246). Diatomites also occur at various levels in Japan, where they are being intensively studied by my friends at the University of Nagoya.

Glacial deposits

Pleistocene glacial deposits on land are a special case and were certainly 'catastrophic' compared with most of the rest of the sedimentary record. The extensive deposits of tills and associated deposits in the Pleistocene are to most geologists just a nuisance, because they cover up the 'real' rocks. I sometimes say that I concentrate on the area between the southern limit of the glacials and the northern limit of the laterites, which are produced by hot, humid weathering and are equally responsible for hiding the real rocks in the tropics. Fortunately this also happens to be the best region for wine production!

Again there is little doubt that tills, outwash gravels and the rest were rather brief phenomena in geological terms. Denmark is the best country I know in Europe for studying such deposits. On the island of Møn one sees the deposits of one glacial lobe overturning the deposits of an earlier glacial advance and at Kirkebjerg one even sees overturned graded bedding in meltwater deposits. Esker deposits (formed by streams flowing

Figure 4.11   Deep weathering of Archaean Morton Gneiss below transgressive marine Cretaceous, near Redwood Falls, Minnesota, USA. Photo DVA.

Figure 4.12 Miocene kaolin being worked in adits with a thin red band below the left adit marking a very brief extension of a river flood-plain, Pugu Hills, Tanzania. Photo DVA.

beneath the ice) are seen at many places such as Stenskoven and kames (mounds of gravel left at the end of such streams) are well seen at Kuls Bjerge. Morainic ridges are everywhere as are erratic blocks. On the quay-side at Frederikshavn there is a huge block of larvikite carried by the ice from the Oslo fjord (Figure 4.13) and Norwegian erratics are known as far south as London. In my beloved Jura Mountains there are erratic blocks of nasty metamorphic rocks coming from the Alps resting on the beautiful local limestones. Lyell figured such blocks on Shetland (1990 facsimile edition, p. 260) but would not have dreamed of postulating anything as 'catastrophic' as a major glaciation. So far as tectonics are concerned, I sometimes tell tectonicians that the best structures to be seen in Britain are not in the Precambrian rocks of the Scottish highlands where they all gather, but in the Pleistocene rocks of gentle Norfolk in the east of England. There, in the cliffs around Overstrand, for example, one can see Upper Cretaceous Chalk and earlier Pleistocene deposits overturned by the onslaught of the ice and at West Runton a great slab of Chalk has been prised up within the till (or 'boulder clay' as we still quaintly call it). The same can be seen in the Island of Møn in Denmark. Ice scratches, actually produced by clasts carried within the ice, are commonly seen in Scotland and first brought the 'Ice Age' home to this country, when the great Swiss scientist Louis Agassiz pointed to

Figure 4.13   Glacial erratic of larvikite from Norway on the quay-side at Frederikshavn in Denmark, with my wife Renée as scale. Photo DVA.

a rock in Edinburgh and said 'That is the work of land ice' (Agassiz, 1886). Agassiz knew well the effects of ice in his native Alps. I often think that the recognition by Penck and Bruckner of four distinct glaciations in the Alps led Quaternary specialists subconsciously to expect four distinct episodes everywhere, but this was not always justified. Certainly there was more than one such catastrophe, as shown for example by the two directions of ice scratches seen in the Fakse Quarry in Denmark (Figure 4.14) and by the different clast contents and orientations in the tills of eastern England.

Nevertheless, four distinct glaciations are clearly recognizable in North America and have been studied in great detail. As always happens, more and more careful studies revealed more and more complications, with minor oscillations of the ice going backwards and forwards. Driving across the endless plains of the Mid-West, I was always impressed by the enthusiasm of local specialists pointing out some barely perceptible ridge in the landscape as some named terminal moraine. As many as 20 distinct glaciations have been recognized in the South Island of New Zealand.

Raised beaches and their deposits are seen all round the coasts of Europe where the sea-level fell when so much water was tied up in the ice-sheets. Again they are often only seen with the eye of faith, but the one that impressed me most was a North American one up the coast of California on which runs California Highway 1 (Figure 4.15). On some coasts there are whole staircases of such raised beaches, each clearly representing a brief halt (in geological terms) in changing sea-levels. One that impressed me very much was north of Agadir in Morocco. Driving south towards the village of Tamri one comes to a flat area with sand-

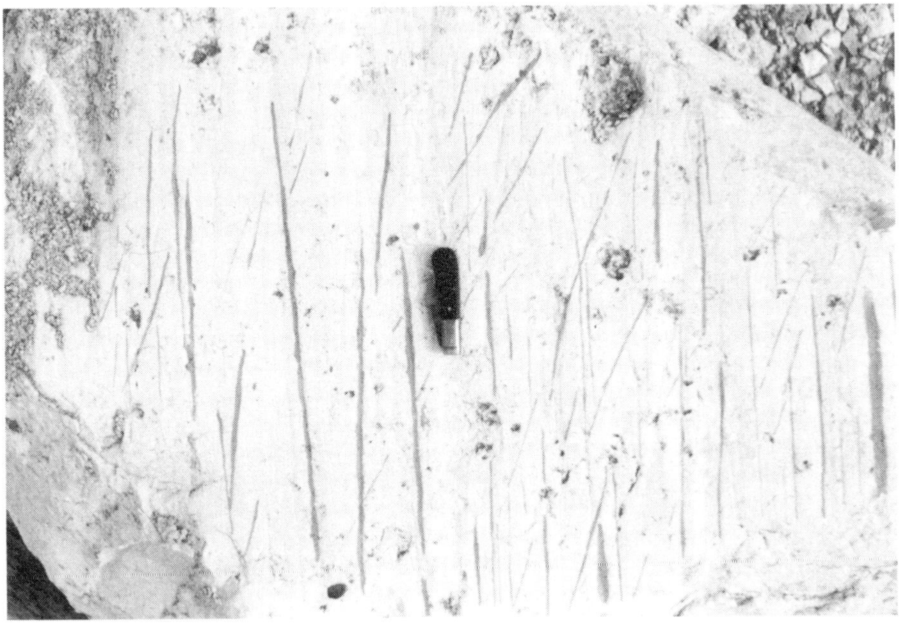

Figure 4.14   Ice scratches in two directions in the Chalk of Fakse Quarry in Denmark, with my lost penknife as scale (perhaps that is where I left it!). Photo DVA.

Figure 4.15   Raised beach platform about 10 m above sea-level, on California Highway 1, south of Carmel, USA. Photo DVA.

dunes as on a beach and one half expects to see the sea round the next corner. One does see the sea, but it is about a thousand feet (305 m) below. The beach deposits at the top of the cliff here are very similar to those on the modern beach below and even contain what look to my eyes like the same species of molluscs. One presumes therefore, that the change in sea-level here was very rapid.

The Pleistocene is the geological period closest to us in time, so we should know it better than any of the others. When one thinks of the comings and goings of the ice, the changes in sea-levels, the changes in geography and the changes in the fauna and flora (including mankind with all our disasters), one must suspect that the other more distant geological periods were equally catastrophic!

Figure 4.16   A 'dropstone' associated with the Late Precambrian Varanger Tillite, 30 km west of Rustefjelbma, Arctic Norway. Photo DVA.

There were, of course, earlier glaciations. I have seen the Late Precambrian Varanger tillite at Vestertana near Rustefjelbma in Arctic Norway. Besides the tillite there are occasional isolated boulders dropped from the ice-sheet in fine-grained sediment (Figure 4.16). This tillite extends all the way down to Scotland and was evidently a major, if brief, event. I have seen a similar (though less photogenic) Precambrian tillite by the Yangste River in southern China and nearby Mount Lushan is capped with Pleistocene till. Late Proterozoic glacial deposits have been described in many parts of the world including, for example, Brazil. Late Palaeozoic glaciations were similarly widespread. Thus I have seen a Permian tillite on Bren Spur in Kashmir (Figure 4.17). Harold Wanless (1960) provided evidence for multiple Late Palaeozoic glaciations in Australia.

Figure 4.17   Boulder in Permian tillite, Bren Spur, Kashmir, India. Photo DVA.

I have not seen the Ordovician glacials of North Africa, but I do not need convincing that though such 'catastrophic' glaciations were rare events, they did happen several times in earth history, often in repeated short-lived pulses. Their cause is still disputed, but I am content to say here that they happened and that their effects were catastrophic.

The evidence clearly indicates that, for most of that history, the earth was a milder and more pleasant place to live than it is at the present day. We are living in the aftermath of the Pleistocene catastrophe and, even without the concern of the ecologists about man burning fossil fuels, there is no doubt – on purely geological grounds – that global warming must come. With it there will be melting of the polar ice sheets, renewed marine transgressions across the continents and the renewed deposition of platform carbonates. All this will be discussed in Chapter 7.

I conclude this chapter therefore with the comment that I find nothing in earth history more convincing of the catastrophic/episodic nature of the record than that of ancient terrestrial deposits.

5
Modern shorelines

The most famous shoreline of my generation was that of the beaches of
Normandy and the 'D-day' landings of 6 June 1944. My first involve-
ment with Recent sediments (other than the sand-castles and dams of
my childhood) was in the trials that preceded the D-day landings.

There had been the disastrous raid by British and Canadians on Dieppe
in 1941 and the Americans had carried out an ill-fated landing exercise
in south Devon in which many lives were lost. So a national appeal
went out in Britain for post-cards or holiday snap-shots of every
imaginable invasion beach in western Europe. All possible information
was needed before the Allies launched the greatest amphibious operation in history.

Then two British commandos were put ashore, long before D-day, to
sample the Normandy beaches with hand-augers. They had to hide in
the shadow of a break-water while a German sentry sat on the end of it
smoking his pipe. As one of them swam back to the craft from which
he had been secretly landed, the auger slipped from his belt. When he
got back to Britain and it was found that the auger was missing, there
was an almighty 'flap' as it was called and only those who have been in
the forces know what that means. Had the auger been dropped below or
above low tide level? If the latter, then the whole secret of the Normandy
landings would have been revealed, with resulting disaster for our forces.
For a few days I understand there was a mad scheme to fly Mosquito
aircraft along every possible invasion beach in western Europe, dropping
augers. I don't think they had enough augers! Anyway the project was
abandoned and everyone prayed that the secret was safe.

I was not involved in that side of things nor in the actual landings
(through no fault of my own, though I often think of Henry V's speech
before Agincourt about 'gentlemen in England, now a-bed'). I was in
an experimental tank battalion and I was only engaged in the trials in
Britain on the closest approach we could find to the Normandy beaches
(Ager 1985a). The beaches chosen for our experiments were those at
Brancaster in north Norfolk (Figure 5.1). As in Normandy, less than a
metre of sand here rests on peat and sticky grey clay. Such a thin development of Recent sandy sediment appears to be a very widespread phenom-

Figure 5.1   Brancaster Beach, Norfolk, England seen in more peaceful times with the Golf House (which was our battalion headquarters) in the background and the sticky grey clay showing up just below the sand in the foreground. Photo DVA.

enon in certain circumstances and indicates the ephemeral nature of modern clastic shorelines.

My unit was first asked to find out how many tanks could be driven across the Brancaster beach before they bogged down. The answer was one! We lost three tanks that were thoroughly stuck when the tide came in (Figure 5.2). We then started to experiment with artificial road-ways. First we tried interlocking metal slats such as were used for temporary aircraft landing strips. But we soon found that the metal toggles were invariably bent and rusted and what might have fitted together easily on a nice dry grassy airfield were hellishly difficult to articulate on a wet and irregular beach. Then we tried burlap matting strengthened with scaffolding poles. I spent many hours in March 1944 struggling in an icy North Sea unrolling great swiss-rolls of this stuff. It was meant to drop down from a great bobbin mounted on the front of a tank and to unroll as it passed under the tank's tracks, but military life is never as simple as that.

The operation was not made any easier, either in the trials or on the day itself, by the other specialist tanks we had beating the beach with chains to explode mines, which made the beach even more unstable. There were many other tank devices designed to clear or span the obstacles that the Germans had so unkindly provided to delay our progress. We made copies of these and experimented with ways of clearing them. There was 'Element A', 'Element B' and 'Element C'. On my 21st birthday we laid on a great demonstration for General Montgomery, to

Figure 5.2   Churchill tank bogged down on the beach at Brancaster, Norfolk, England in the trials in preparation for the landings in Normandy in 1944. From *Vanguard of Victory* (1984, Plate 38) by David Fletcher. Reproduced by permission of Her Majesty's Stationery Office, London.

show him how we would deal with the Normandy beaches. My particular tank had to cope with one of the 'Element Cs' (immediately christened 'Elephant's knees' by the irreverent soldiery). This was little more than a large iron gate-way propped up by heavy metal struts. We decided that the easiest way to deal with this was simply to charge it at full speed, which in our Churchill tanks on the hardest of road-ways was only about 24 miles (*c.* 36 km) per hour. But we could not achieve that immense speed on the sticky sediment and we moved sluggishly up the beach to strike the obstacle with a dull thud. So it was that I spent most of my birthday perched at a ridiculous angle and not able to go forward or back. Meanwhile another tank came up alongside and tried another method. This was to lob a small dust-bin (trash-can in American) full of explosives from a stubby mortar on its front with the fine old mediaeval name of 'petard'. As Hamlet said we were duly hoisted! It did not work but (unknown to us), it blew open the lid of the storage bin on the back of our turret so that it came to rest on our radio aerials, shorting them and putting us out of communication with our commanding officer. As a result we were falsely written off well before D-day. All these reminiscences of a best forgotten past may seem irrelevant to geology, unless it was a problem for the heavier dinosaurs. I only trouble the reader with them to emphasize the unconsolidated nature of a certain kind of modern shoreline. When the actual landings came, mainly lighter vehicles were used, but many tanks adapted for clearing mines (Figure 5.3) were, as expected, bogged down in the first dash across the beaches, though our road-ways and beach-clearing devices (and our knowledge of the underlying geology) undoubtedly saved many Allied lives. My brother, who landed his heavy tank in Normandy a few days later, vaguely remembers passing hurriedly over a metal track-way, which by then had been properly laid in appropriate places.*

Figure 5.3 'Crab' tank equipped with flails for clearing mines bogged down on a Normandy beach. From *Vanguard of Victory* (1984, Plate 39) by David Fletcher. Reproduced by permission of Her Majesty's Stationery Office, London.

We had offered our devices to the Americans, but they had brushed them aside with the bold statement 'Oh, we'll leave it to the marines'. I always thought that was why they suffered such dreadful casualties on Omaha Beach that first day, but it was not until the 40th anniversary of D-day, in 1984, that I heard that stated officially.

It should be noted that both the Normandy beaches and those in Norfolk are, by their geography, comparatively protected from the worst effects of the elements. Not far to the west of those well-remembered

---

* He belongs to the Normandy Veterans Association and the president of his branch is, curiously enough, a Captain D. Day!

Brancaster beaches is the great and even more protected embayment of the Wash with its ephemeral sediments, well described by my friend Graham Evans (1965) and later discussed by myself (Ager 1981a, p. 48). Here the sediments seem to come and go with the seasons. The silts and muds come in during the summer, only to be carried out again in the winter. Certainly there is a well-stratified succession in the walls of the creeks crossing the mudflats (Ager, *op. cit.*, Plate 5.1). These presumably represent annual layers, but they are constantly being eroded as the creeks migrate. Nothing is permanent.

It was here, early in the thirteenth century, that King John of England lost his baggage train with all his valuables including, by tradition, most of the Crown Jewels. They sank into the mud in conditions rather similar to those of the tanks in nearby Norfolk seven centuries later. Of course he, like the Americans, lacked the advice and experience of my experimental battalion! So it may be said that King John's inadequate knowledge of shoreline geology went with his other inadequacies as a mediaeval monarch, which led to the signing of *Magna Carta* as the foundation of British and American laws and liberties.

But why did we have these problems? Why are the beaches of north Norfolk and Normandy like that? If one looks at a map it is immediately obvious that they are similarly placed geographically, with a land-mass protecting them on their western sides. In the case of Norfolk and the Wash this is the main mass of northern England. In the case of Normandy it is the Cherbourg Peninsula (Cotentin), which was vital to the D-day strategy of rapidly gaining possession of a large port (i.e. Cherbourg). The comparatively protected nature of the coast also made possible the building of the prefabricated 'Mulberry harbour' at Arromanches, though this was damaged soon afterwards by an unexpected storm coming from the wrong direction.

Another well-protected beach which is now particularly well known to me is that in the tightly enclosed Swansea Bay outside my study window at Mumbles. At night-time, the ring of lights makes it appear that Dylan Thomas's 'mumbling bay' has no outlet to the sea at all. I compare it with the Bay of Naples with the red glow above the Port Talbot steel works, on the far side, playing the part of Vesuvius.

As my department knew to its cost, Swansea Bay is equally treacherous when it comes to losing vehicles. A research student, who shall be nameless, foolishly took our departmental Land Rover down on to the damp outermost part of the beach at low tide and got it thoroughly bogged down before the tide came in again. Here again the thin and discontinuous sand rests on sticky grey clay and peat. Scour action was such that the Land Rover was almost down to roof level in the sediment in just one passage of the sea (Figure 5.4). This is worth remembering when one comes to consider matters such as the burial of fossils in the strata as discussed by Art Boucot (1953) and by Ferguson (1978). These

Figure 5.4   Land Rover bogged down on Swansea beach, South Wales and almost buried after one passage of the tide. Photo D. J. Lancey.

processes happen very quickly, even if the actual sedimentation rate is very slow, as can be seen with shells on modern beaches. In places no sediment seems to be accumulating at all, as on the beaches just discussed. In Swansea Bay, for example, hardly any sediment seems to have accumulated in 6000 years, from the evidence of a submerged forest which is frequently exposed (Ager 1976). Sediment only accumulates when ill-advised and unscrupulous 'developers' interfere with the fundamental geography of the bay.

Swansea Bay has the second highest tidal range in the world and as a result it has a very wide sandy and muddy beach. To the right from my study window is Oystermouth Castle* probably first built about 1100 AD. It was built like the other Norman castles of Wales to hold down the local population (who repeatedly destroyed it). It was designed, like the others, to be supplied and reached by sea. So was the even older nearby fortified tower of All Saints Church, Oystermouth. That takes us back nearly a thousand years with the shoreline still in the same position, apart from the made ground of debris and discarded oyster shells (from which the village got its name and of which Prime Minister Gladstone was particularly fond). On this and later structures

* which we are now told to call Castell Ystumllwynarth, though no-one in Swansea seems to know the place!

the village has extended in a seaward direction, though all the locals hope that a further huge and greedy scheme of reclamation will not go ahead. Returning to the natural form of the coast-line, however, if one goes farther on beyond the village, one comes to Mumbles Head where the Romans worked iron-ore in the Lower Carboniferous limestone (or presumably conscripted the Welsh to do the hard work). They too presumably needed the sea nearby to carry away the ore. That takes us back nearly 2000 years. If I dig my garden in Mumbles I soon find myself in hard contact with solid Carboniferous rocks. These are exposed on the foreshore below my local 'pub' at West Cross. But if I cross the road to dig in the beach opposite, I quickly find grey clay and peat with abundant organic material (and methane gas). It is the same all round the bay. Opposite the University College of Swansea (which also faces the sea) the 'submerged forest' is well exposed with tree stumps and red deer antlers dating from Neolithic times (Ager 1976).

Thus the coast-line has stayed the same here in Swansea Bay for nearly 1000 years (for the Normans), nearly 2000 years (for the Romans) and for about 6000 years for the submerged forest. What is more, immediately to the west is the beautiful Gower Peninsula with its limestone cliffs and famous prehistoric caves. My friend Mike Bridges has shown (1987) that there has been hardly any erosion of these cliffs in 10000 years. Hardly any sediment has accumulated in Swansea Bay since the sea drowned the forest some 6000 years ago. Therefore, though our shoreline seems unchanging, it is in fact, geologically ephemeral, since there will be no record of it for the geologist of the future.

So what do the sandy beaches of Normandy, Norfolk and Swansea have in common? Firstly, they have only been there a very short time in geological terms and are little more than veneers on older sediments. Secondly, due to their protected positions, the sands were deposited in low energy conditions and are poorly consolidated, which makes them unsuitable for wheeled or tracked traffic.

Rarely do I see a rough sea from my windows, rarely was it not possible to launch my sailing dinghy (the *Rhynchonella*) into the peaceful waters of Swansea Bay. However, I do not have to go far, just beyond the quaintly named Mumbles Head (which I see from my study window) to enjoy the full onslaught of the Atlantic rollers and the waves crashing on the limestone cliffs. Not very far again, across the estuary north of Gower and beyond the protection of its headlands are the Pendine Sands in Carmarthen Bay which are exposed to the full force of the prevailing south-westerly winds, coming in from the open Atlantic. These sands were used for land speed record breaking attempts in the 1920s. They are still well-known in the summer months for the large numbers of holiday-makers' cars which are driven out and parked actually on the beach (Figure 5.5). The same happens at Weston-super-Mare across the Bristol Channel, in my native England, which is similarly exposed.

Figure 5.5   Pendine Sands, south-west of Carmarthen, South Wales, showing many vehicles driven on to the firm sand. Photo DVA.

Down the west side of the northern peninsula of the North Island of New Zealand is 'Ninety Mile Beach', which happens to be 65 miles long (not only Texans exaggerate). This is exposed to the fury of the Tasman Sea and is the regular route of tourist coaches which meander wildly, but safely, across the wide firm sandy beach while the driver (with typical New Zealand humour) pretends to read a newspaper (Figure 5.6). (The same driver points out Cemetery Road with the sign 'No exit'). Halfway along the beach, at The Bluff, one sees solid rock, in this case pillow lavas, not far beneath the surface.

Both the Pendine Sands and Ninety Mile Beach are suited to carry heavy vehicles which would certainly get stuck on those invasion beaches of northern France. They are obviously suited to this treatment because they are exposed to high energy conditions and their sands are well consolidated. This may give us a clue to the environment of ancient beaches, but there is no evidence that any of these is anything but ephemeral in geological terms. The sand is nothing more than a thin manifestation of a recent rise in sea level. They are examples of the episodicity of the geological record, which I will emphasize repeatedly in this book. They are like the *stasis* and *punctuation* of the fossil record, the periodicity of volcanic activity, the rushes of the occasional turbidity current. They are in my much quoted metaphor of the life of a soldier: 'long periods of boredom separated by short periods of terror'.

Figure 5.6 View from inside of a large coach driving on the firm sand of 'Ninety Mile Beach', with the Tasman Sea on the right. North Island, New Zealand. Photo DVA.

In my travels around the world as a hard-working geologist, I am sometimes slightly irritated at being referred to as a 'holiday-maker' (though I don't refuse the cheaper tourist fares). If I *had* been a holiday-maker my ideas of shorelines would probably have been quite different from those I have gained as a geologist. Holiday-makers concentrate on the quartz sand beaches and those of fine coral debris, but I cannot imagine anything more unutterably boring than hours of sun-bathing on such a beach. It is my impression that most of the beaches of the world, especially in the oceanic areas, are made of the dark debris of volcanic rocks. The popular topless beaches of Tahiti (no illustration) are only very local developments due to the presence of offshore coral reefs which give rise to the white stretches, otherwise the beaches are grey due to the weathering of the volcanic rocks of the island. Similarly in Fiji there are the white beaches along the southern 'Coral Coast', but elsewhere there are dark beaches, mangrove swamps and a coastal plain frequently hit by the hurricanes or typhoons, which are discussed in Chapter 9.

Perhaps the most glorious beaches I have seen are those on the east coast of Africa. I think particularly of Kikambula and Malindi in Kenya, where the beaches are dazzling white and riddled with the holes made by burrowing crabs. They are composed entirely of material from the off-shore coral reef, on which the surf roars continuously like an express train and the palm trees at the top of the beach frequently reminded me of Kipling's 'palm grove's droned lament before Levuka's Trade'. I had long wondered where Levuka was (and our geographers couldn't help me) but that takes me back to Fiji, for I discovered there that it was on the island of Ovalau and was the old capital of the Fiji group of islands.

Similarly in Tanzania I remember my little home within a few steps of the Indian Ocean, where the shore-ward side of the beach was stabilized by mangroves. Now that I find the English Channel too cold at my age for bathing and even the Mediterranean hardly bearable, I find the Indian Ocean is gloriously warm, even at six o'clock in the morning, and I have lazed in the water watching the dhows sailing down the Mozambique Channel, steered by a helmsman with one leg nonchalantly over the tiller. Offshore there is Mbudya Island, where the time-oblivious boatman left me to roast on a white beach scattered with debris from the reef which surrounds the island. On the protected shore-ward side I noted that the coral fragments were of delicately branching forms, not suited to the onslaught of the open ocean, but on the opposite, exposed side there were only massive forms. In places the reef had already been transformed into cliffs of coral rock and it was startling for me to notice crabs climbing above my head.

That coral rock was only formed yesterday, geologically speaking, and though my idle reminiscences of past travels may seem irrelevant, they had me thinking about the selective preservation of coastal deposits for the future. Coral reefs and coral sands are obviously soon cemented by calcium carbonate. Quartz sands are presumably less quickly cemented by silica and iron minerals. But if there are so many beach sands in the world today, as I believe, formed of basic volcanic rocks, why do I not know them in lithified form? Is it just a matter of chemistry? Or is it rather that sandy beaches, whatever their composition, do not have a high preservational potential? Back home I remember a geologically ignorant lady complaining bitterly about the 'dirty' beaches of the West Indies, unlike the glorious golden beaches of Gower. She did not realize, of course, that the beaches she had seen on her holiday were not black because of local habits, but because of their derivation from the local volcanic rocks.

We are fortunate in Britain, Shakespeare's 'precious stone set in the silver sea', in that our beaches are largely derived from pre-existing sedimentary rocks, mainly quartz sandstones and cherty limestones, and from plutonic rather than volcanic rocks. I never cease to wonder at the unimaginable quantities of flint pebbles in the shingle beaches of great stretches of north-west Europe, all derived from the Upper Cretaceous Chalk.

Probably the most spectacular shoreline deposit of pebbles in England is Chesil Bank, running out from the Dorset mainland to the so-called island of Portland over a distance of some 25 km, 16 of them separated from the land by the Fleet Lagoon (Figure 5.7). This is almost wholly composed of well-rounded flint pebbles derived from the Chalk to the west. They are so well-graded that it is said that a local fisherman, landing on the beach at night, can tell where he is from the size of the pebbles.

Another obvious factor in the formation of modern beaches is that of tidal range. For all their touristic popularity, the beaches of the Mediterranean are very limited and therefore all the more overcrowded, because of the limited tidal range of this enclosed sea. One may compare them with Caswell Bay on the Gower Peninsula near my home. This is a very wide sandy beach due to the huge tidal range and is very crowded in the summer, due to the convenience of a car park at beach level (though it cannot compete with the Riviera in the sunshine tables). As the tide rises over this funnel-shaped beach, the crowds are forced back into the narrow end of the funnel, forming what a palaeontologist might call a condensed fauna.

Figure 5.7 Chesil Bank seen from the 'island' of Portland, Dorset, south-west England; a long spit of flint pebbles running out from the mainland. Photo DVA.

Even chillier than the Bristol Channel is the Baltic Sea, which has a constricted entrance like the Mediterranean and so a limited tidal range, plus a reduced salinity due to the influx of fresh water from all round. Its beaches are narrow and little used, though the hardy nordic people are not afraid of the cold. The diversity of the marine fauna decreases as one moves up the east coast of Sweden into the Gulf of Bothnia. This is a potential tool for estimating decreases in salinity in ancient seas, as I have suggested for the faunas of the Upper Jurassic in south-west England (Ager 1975a). The common sea-shells, such as the cockles and mussels become progressively smaller as the salinity decreases northwards.

In Japan I was impressed both by the use they made of their surrounding seas for food and how little use they made of them for leisure. I have suggested this to my lone sailor daughter as a possible opening in these days of market economics (much as I dislike the philosophy of the 'grocer's daughter'). Every university in Japan seems to have its Faculty of Fisheries; even a small university such as that of Kagoshima has three ocean-going research vessels and the late Emperor himself was a marine biologist. Leisure on the other hand, seems to consist of art, golf inside great green nets and troops of tourists in temples following little flags. No doubt it is largely a matter of Japanese history and temperament (notably their love of work) but in three months touring those fascinating islands, I saw no marinas for the consumption of excess wealth and nothing I would call a 'family beach'. The holidays that were advertised were either in far away cities or on Pacific islands such as Iwojima with pictures of Japanese bathing beauties reclining on the sunny beaches of tropical coral islands. Again I digress, but I don't think geology should be a dull, single-minded subject and I think this has relevance to what we may expect in the ancient record.

Japan is, of course, exposed to the full fury of the far from pacific Pacific with its typhoons and one must never forget that *tsunami* is a Japanese word. They may be called storm surges but never 'tidal waves'. The only 'pale' beach I saw in tours of the four main islands of Japan was at Shikanoshima ('deer island') near Fukuoka in Kyushu. I went there chiefly to see a Mesozoic granite, which was a strange sight for me, though comparable to the granite of similar age I had seen in the Sierra Nevada of California, on the other side of the Pacific. The pleasant sandy beach at Shikanoshima had presumably been derived from the granite on the more protected western shore of Kyushu, as it has opposite, across the Yellow Sea at Bedaihe, the Chinese holiday resort for Beijing (where the chief entertainment seems to be hiring a camera for an hour and then having your film developed at the kerbside).

Elsewhere in those hospitable Japanese islands, the coast-lines are largely of volcanics and argillaceous sediments. I remember particularly a long scramble, in pouring rain, down to a beach made entirely of purple andesite pebbles at the end of the Oshika Peninsula in north-east Honshu, inappropriately (to my eyes) called the Cobalt Coast.

The battered rocky coast-line of most of Japan makes it particularly suited for growing oysters and the other molluscs and crustacea which form such a delightful part of the delicious Japanese cuisine. One hears much of raw fish (which only horrifies westerners who have not tasted it), but not of the octopus, abalone and other marine animals and plants which are adapted to these rocky coasts. If sand does accumulate here, it seems to be swiftly carried away again.

Most of the Pacific coast lacking the protective coral reefs of the tropics seem to be subject to erosion rather than deposition. One sees this clearly, for example, if one drives up California Highway 1 on the splendid raised beach between Los Angeles and San Francisco (Figure 4.15) and the offshore erosional topography is particularly marked. Thus the submarine canyon off Monterey Bay exceeds the Grand Canyon in dimensions (Figure 8.1). Similarly, if we go back to the gentler scenery of eastern England and go north from the Wash into the East Riding of Yorkshire (which we are now supposed to call Humberside) we find active coastal erosion, with roads disappearing at the edge of crumbling cliffs of Pleistocene till. South too, a great deal of the land of the Angles (East Anglia, heartland of the Anglo-Saxons) has been devoured by the ravenous North Sea. The town of Dunwich was the great centre for this region in mediaeval times. Before the Great Reform Act of 1832, less than a dozen electors remained to send two Members to Parliament. But the last of the old churches fell into the sea in 1919. When I was there in the 1950s, one could still find human bones and coffin handles scattered on the sandy beach.

When I was a student I collected many beautiful fossils from what was probably the finest section for palaeontologists in England – at Barton-on-Sea in Hampshire. It was a marvellous section because the sea was constantly eroding the Eocene clay cliffs with multitudes of exotic molluscs. Here the sea was advancing and with it the narrow sandy beach, but again nothing was being preserved for the future. Unfortunately the local inhabitants most unreasonably objected to their houses falling into the sea, so Neptune is now restrained and this magnificent section is largely hidden behind concrete.

The first English colony in North America was established by Sir Walter Raleigh in 1585 on Roanoke Island off the coast of what is now North Carolina. In Sir Francis Drake's old house near Plymouth, it is recorded that on his return from one of his great voyages (of what amounts to piracy) in 1586, he rescued 103 starving settlers from Roanoke. Not to be deterred, Raleigh (not personally, but through what might be called an agent) established a second colony there. Five years later this second colony (often called the first) had disappeared and their fate remains a mystery to this day according to most authorities (e.g. Dolan & Lins 1985). This could have been another geological catastrophe, since the north side of the island where the colony was probably sited has been steadily eroded over the past four centuries and this erosion continues to the present day.

Therefore, whether the shoreline is stable or advancing, whether the shoreline sediments are firm or treacherous, it seems that these are fleeting things that are rarely preserved for the future. An advancing sea consumes its own sediments and probably the best chance of preserving such a shoreline is when the sea is withdrawing from the land.

Great attempts have been made to quantify the sediment budgets of various coastal areas, but I am not convinced of the success of these efforts. Thus the Geological Society of London published a special set of papers on 'Coastal Margins as Sediment Traps' (Kirby *et al.* 1987). The sediment accumulates, if at all, in great embayments such as the Wash and the Moray Firth (on the east coast of Scotland), but even in these traps it is anything but permanent.

A study of the coastal sediments between Turakirae Head and East-bourne in Wellington Bay, New Zealand (Matthews 1980) showed that sand and gravel rapidly disappears from exposed beaches here to be redeposited (at great inconvenience) in the harbour. Even during the five very pleasant weeks I spent there in 1986, I saw recently deposited beach sands being rapidly eroded from just across the road from our temporary home. On the opposite side of Wellington Bay is Oriental Bay (named not from its position – it is on the north side of the great embayment – but from the name of the ship that landed colonists there in 1840). It is very popular with the inhabitants of New Zealand's capital, both because of its beauty and of its possession of the only sandy beach close to the city centre. But the sand is a problem (Lewis, Pickrill & Carter 1981). It has to be artificially replenished for the benefit of the citizens, but Nature does not cooperate and steadily removes the sand again.

It may seem that I have contradicted myself (not for the first time) in that in one breath I have said that coastlines are static and in the next that they are ephemeral. But of course I speak in human terms. My point is episodicity. Nothing may happen for long periods of time, far beyond human experience, and then perhaps everything is swept away and redeposited in one day. What is clear, at least to me, is that the shoreline and its sediments are not preserved for the stratigrapher of the future. One needs very special circumstances for that to happen.

Apart from the gradual building out of deltas, with the sediment derived from erosion inland and blown sand moving inland as dunes, nothing seems to be building up along the coastlines or on most of the nearshore shelf. We see sandy veneers (as in Normandy) on older sediments. David Carter told me (personal communication in the 1960s) that the preliminary studies for the Channel Tunnel (between England and France) showed the meerest veneer of sediment on top of the Upper Cretaceous Chalk. Offshore sand-banks are constantly in motion (see Chapter 8). We do not see successions.

No doubt all this is a reflection of the times in which we live and the rise in sea-level in Holocene times. Perhaps we can only expect shorelines and shoreline deposits to be preserved in a regressive situation as the seas withdraw once more in the next glaciation.

6

Ancient shorelines

I largely blame the more eccentric of my two eccentric grandfathers for my becoming a geologist. It is not that I knew him, he died when I was one year old, but as a child I devoured voraciously his large and rather odd collection of books, many of which were concerned with the margins of science. One which I still retain is *Scepticism and Geology* by an anonymous gentleman who called himself 'Verifier' (1877). It is essentially an attack on the principles of uniformitarianism, as expounded by Charles Lyell. Though I am all for healthy scepticism, I do not accept the main emphasis of this book which is that there has been little or no change in the history of the Earth. I would not have it thought that it is because of this strange book that I find it difficult to accept the classic views of Lyell.

To return to my eccentric grandfather, however, he had annotated the book in his own distinctive handwriting. One such addition marks a passage on sedimentation and refers the reader to Jeremiah 5, 22. It is not my usual custom to quote holy writ in my geology, but the text referred to is highly relevant to this chapter. It may be said to be a proclamation of the permanence of clastic shorelines. It reads:

Fear ye not me? saith the Lord: will ye not tremble at my presence, which have placed the sand for the bound of the sea by a perpetual decree, that it cannot pass it: and though the waves thereof toss themselves, yet can they not prevail: though they roar, yet can they not pass over it?

Of course, in reading this we must think of the biblical lands at the eastern end of the Mediterranean, with its very small tidal range. This text hardly tallies, however, with the transgression of Noah's flood in Genesis or with the rapid regression and transgression of the Red Sea in Exodus. But it is not my business here to discuss the many contradictions in that strange book. Nevertheless, it may be said that transgressions can be very rapid and one knows that disastrous tsunamis or storm surges are commonly preceded by a withdrawal of the sea, which might have been long enough to allow the Israelites to make their escape from Egypt if they were very quick! Nevertheless, the above does illustrate an attitude of permanence which pervades much of the geological literature, even though we are constantly preaching the doctrine of continuous change.

In that connection, it is worth considering the restricted inland seas of the past, such as the so-called 'Niobrara Sea' of Late Cretaceous times in South Dakota, Nebraska, Kansas, Texas and parts of neighbouring states and the Late Permian 'Zechstein Sea' of northern Europe. Stearn, Carroll & Clark commented (1979, p. 335) on the beautifully preserved vertebrate fossils that are found in the former, including birds, swimming reptiles and flying reptiles, all of which still have unbroken and articulated skeletons and are interpreted by the authors as indicating that 'the waters of the Niobrara Sea were untroubled by waves and currents'. How did these inland seas compare with the broad Tethys or the early Atlantic? The presumption that sandy beaches always remain in the same place is clearly not true. This very evening as I write this (in December 1991), the local paper reports that 55 million tonnes of sand have been removed from Swansea Bay in the last 52 years and that another 59 million tonnes could disappear in the next 20 years. This is largely due to the lack of foresight by the local planners, but also to their ignoring the advice from the local geology department! The possibility of resultant flooding makes me a little concerned about the vulnerability of my sea-facing house. As previously mentioned the tidal range in Swansea Bay is up to about 13 m (the second highest in the world I am told). This is very different from the one metre or so in the Mediterranean. Perhaps all this should go in the previous chapter but it happens to be apposite in this context.

Going back to the Niobrara Sea, the Americans recognize a narrow, sandstone shoreline belt between the marine and the terrestrial. This is the Dakota Sandstone and is transgressive at the base of the Upper Cretaceous. The only specimen I have of it has a perfectly preserved birch-like leaf and is a reminder that this was the time of the first flowering plants and that trees are more likely to be preserved than herbaceous plants (see Chapter 2). This transgressive sandy facies is likely to be diachronous, if only on a minor scale. The sandy beach, if that is what it was, did not stay in any one place for long.

The best demonstration I know, of a diachronous marginal sandy facies is in the intensely studied Jurassic rocks of southern England. If one starts at the glorious ribbed sandstone cliffs near Bridport in Dorset (Figure 6.1), one sees the Bridport Sands which range from the 'Upper Lias' (Toarcian) to the Lower 'Inferior Oolite' (Aalenian) in age; I ask my foreign friends to excuse the quaint old English stratigraphical terms. The next illustration (Figure 6.2) demonstrates the diachronism. The horizontal lines on the diagram represent successive ammonite zones. If one goes farther north into Somerset, exactly similar deposits change their name to the Yeovil Sands, which are wholly in the uppermost Toarcian. North again, we cross the Mendip Hills (where the Middle

Figure 6.1   Bridport Sands (Jurassic), Burton Cliff, Dorset, south-west England.
Photo DVA.

Jurassic rests directly on Dinantian/Mississippian) and find the Midford
Sands, lower down in the ammonite zones of the Toarcian and finally
we see them in my beloved Cotswold Hills of Gloucestershire, farther
down still and overlain by the 'Cephalopod Bed', full of Late Toarcian
ammonites and belemnites. In other words there was a belt of coastal
sands which moved south with time and did not remain idle in any one
place.

   If one looks at palaeogeographical reconstructions one often sees the
presumption that coastlines remained in the same place for countless
aeons. My adopted homeland of Wales, for example, has been unkindly
called 'St George's Land', after the patron saint of England (and of
Portugal, Bulgaria and Russia for that matter) though that saint has now
been desanctified and 'St David's Land' (named after the Welsh patron
saint) would have been more in keeping with Welsh sensitivity. One
would think from the maps that the waves battered the same rocky
coastline of this land for hundreds of millions of years without effecting
any changes. One would think from the maps that the beaches up which
the amphibians first waddled out of the sea were still there for the ageing
dinosaurs to enjoy the last of the Cretaceous sunshine and that they are
still there today for the young of *Homo sapiens cwmryi* to practise the
Welsh religion of rugby.

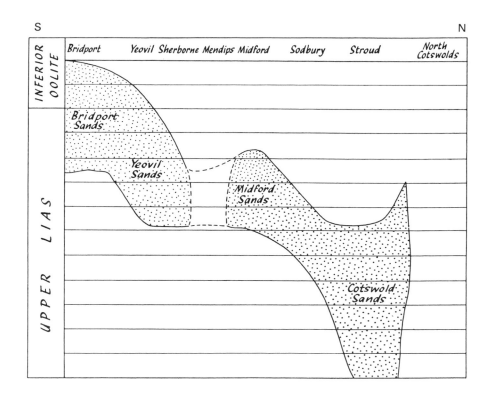

Figure 6.2   Diagram showing the diachronism of the Bridport Sands and their lithological equivalents in southern England. The 'Upper Lias' = Toarcian and the Lower 'Inferior Oolite' = Aalenian in the international terminology. The horizontal lines represent ammonite zones. Drawn by Miss M. E. Pugh.

This cannot be. Coastlines are constantly changing and as the coastlines move, so do the beaches which mark their margins. Marine transgressions are by definition transgressive. As the sea advances across the land, then the unconsolidated, uncemented sediments of the shoreline are likely to be whipped up and carried forward over and over again. If the coastline remains fixed in one place, then nothing changes, nothing accumulates for the future, as in the examples I gave in Chapter 5 from the present day. It is perhaps, only in times of regression that the coastline sediments stand a good chance of being preserved, but then subaerial erosion comes into play and the coastal sediments are likely to be transported into deeper water.

In all my geological wanderings around Europe, North America, Africa, Asia and the South Pacific, I have been repeatedly impressed by the persistence of particular facies at particular moments in geological time. This applies especially to shallow water carbonate deposits (see Chapter 7) and to continental sediments such as 'coal measures' and 'red beds' (Ager 1981a, Chapter 1). It appears to be very rare, however, for shoreline arenaceous deposits of a particular age to be preserved over a wide area. The one good exception I know is the Grès armoricain of north-west France, for example in the cliffs of the Crozon Peninsula in Brittany and forming the escarpment on which was built the castle of the Dukes of Normandy at Falaise. This Early Ordovician quartzite may extend northwards into the west of England as the Stiperstones Quartzite of Shropshire. In the other direction it seems to extend via the Massif Central to northern Spain, where it is seen as the Barrios Quartzite which forms such a marked feature in the Cantabrian mountains (and was important in Spanish military history). It is seen again on the south side of the Iberian Meseta. Beyond that it may well continue into north-west Africa. Everywhere it is a fine-grained quartzite, usually pure white in colour, but with some notable exceptions (such as the 'liver-coloured' quartzites of Brittany, which are found as derived pebbles in the basal Triassic of southern England).

Comparable perhaps is the basal Cambrian quartzite which, in the north-west highlands of Scotland, looks remarkably like rocks of the same age in the Canadian Rockies, for example on Mount Eisenhower (or is it again Castle Mountain?). Even the trace fossils, known as 'pipe rock', are the same (though they are seen with much less effort as loose boulders on the shore of scenic Lake Louise in the Canadian Rockies).

On both occasions no doubt, at the beginning of the Cambrian and at the beginning of the Ordovician, something funny was going on. Certainly there were very widespread and probably rapid marine transgressions after long breaks in the marine record. In these cases, in spite of what I said above, the first deposits of marine transgressions do seem to have been preserved, though perhaps we have lost the pebble beaches. Even the latter have been preserved locally in the Ordovician at Builth in Wales (Jones & Pugh 1949); see also Ager (1981a, p. 43 and Plate 4.4). In the transgressive Late Llandovery (Early Silurian) there are pebbly beaches preserved around the former islands of Precambrian rocks that still stand as the hills of the Longmynd, Breidden, Caer Caradoc and the Malverns in the Welsh Borderland of England (Whittard 1932). Going higher up the stratigraphical column we have the conglomeratic Sutton Stone at the base of the Lower Jurassic in the Vale of Glamorgan in South Wales. This has commonly been interpreted as the first pebbly deposit of a transgressive sea. I reinterpreted it as a storm deposit (Ager 1986b) or what I called a 'Tuesday afternoon deposit'. (see Chapter 9). In other words it was a very fleeting phenomenon (though I don't know why Tuesday afternoon sounds funnier than any other time of the week). It will be discussed again in Chapter 9. My ideas were criticized by the Geological Survey establishment, but my head is not bowed and they were cited with evident approval by Johnson (1988).

In the latter paper, Johnson discusses the whole question of 'Why are ancient rocky shores so uncommon?' This is true and surprising when one remembers that 33% of modern shorelines are of this nature. The answer partly relates to the special conditions of today (see Chapter 12) and largely to the very special conditions needed for their preservation. One of the best examples I have seen was on the island of Ivø in a lake north-east of Kristianstad in southern Sweden, where Upper Cretaceous (Campanian) sediments rest on Precambrian metamorphic rocks and contain boulders of gneiss from the latter, encrusted with a rich and diverse fauna including oysters, serpulids, bryozoans, cemented brachiopods and other forms (Surlyk & Christensen 1974). All these indicate a littoral environment. Another good example of a different kind was near Cernotin in Czechoslovakia, where Middle Miocene (Helvetian) rests unconformably on Devonian limestones and the latter are bored by the bivalve *Lithophaga*, some of which are still preserved in the holes. Markes Johnson informed me (personal communication 1988) that he had visited the locality I had described near Kolin in Czechoslovakia (1981a) and had found encrusted surfaces at other localities in the neighbourhood.

Bored surfaces encrusted with oysters are in fact quite common in the stratigraphical record, at least above the level at which oysters first appear. I know them best in the Middle Jurassic (Aalenian–Bajocian) in my favourite part of England, the Cotswold Hills (Ager 1955, Ager *et al.* 1973) and in the Lower Cretaceous (Hauterivian–?Barremian) in my favourite part of France, the southern Jura mountains (Ager & Evamy 1963, p. 341). I have seen them in the Lower Cretaceous of Morocco and of Tunisia. It was the marine borings in Roman pillars at Pozzuoli near Naples that Lyell used as the great symbol of uniformitarianism, to be illustrated in the frontispiece of his *Principles of Geology* in all its editions. I have argued, however, (Ager 1989) that far from illustrating uniformitarianism, the pillars show the reverse, which is why I have chosen them also for my frontispiece. They show a very rapid change in sea-level, since below the famous borings, which are about three metres above the present high tide level, there are about two metres with no borings at all. I apologise for quoting myself so much, but it is in line with my motto mentioned previously. I sometimes think that Ager is almost the only geologist with whom I agree! I tend to get somewhat cynical in geology and only to believe what I have seen with my own eyes. Hard surfaces are often encrusted with other molluscs such as inoceramids, for example in the Late Cretaceous of the Great Plains and southern Rockies in the States (Hattin 1986).

We know today that local geography can affect the tidal range, as in the North Island of New Zealand, with tides sweeping round from both the Pacific Ocean and the Tasman Sea. We can even see it on a small scale in southern England, where the tides sweeping round both sides of the Isle of Wight mean that the sea at Bournemouth, for example, never retreats very far. Tidal range in ancient times may be reflected in the depth of the vertically burrowing trace fossils, which are of particular interest to me. My favourite rocks – the Jurassic – are sadly lacking at the surface in south-east England, but we will shortly only have to take the Channel Tunnel to see, near its French exit, magnificent sections in Jurassic clastic shorelines, with abundant trace fossils, along the coast between Wissant and Boulogne. These were described by Peigi Wallace and myself (Ager & Wallace 1966a&b, 1971). The sections include several conglomerates (which are very strange to British Jurassic-trained eyes) because of the proximity of the Brabant Massif at that time. There are also several shallow water sandstone formations and this area was clearly, in Late Jurassic times in a regressive situation, as the succession passes up into freshwater tufas (and in underground south-east England into evaporites) of the 'Purbeckian' facies at the top of the system. Above that are the fluviatile, lacustrine and paludal deposits of the Early Cretaceous 'Wealden'.

Under these special circumstances, Upper Jurassic formations such as the Grès de Chatillon, the Grès de la Creche and the Grès des Oies show clear evidence of intertidal conditions, including trace fossils such as *Diplocraterion*. The topmost of the above sandstone formations, the Grès des Oies, is immediately followed by freshwater deposits and is noteworthy for its quicksand and 'sand volcano' structures. It would certainly have been a very unsuitable beach for landing tanks if such things had existed in peaceful Jurassic times! (see Chapter 5). It is worth noting that the sands are thin and only represent brief episodes of clastic deposition as the sea regressed from this area. This happened three times just in this marginal area as the sea oscillated to and fro towards the end of the long and glorious history of the Jurassic Period.

Not very far from here, but of much greater age, is the outcrop of the Grès de Ste Godeleine of the Ferques inlier. These beds were also mentioned in Ager & Wallace (1966a&b) and were described in detail by Peigi Wallace (1969). They are Famennian (latest Devonian) in age and show many features, including trace fossils and quicksand structures, comparable to those in the Jurassic sandstones mentioned above. Also, from the formation's stratigraphical position and the presence of supposedly freshwater bivalves, the Grès de Ste Godeleine clearly represents a complex, but short-lived clastic shoreline facies during a major regression at the Frasnian–Famennian boundary.

The presence of *Diplocraterion* in both the Jurassic and the Devonian sandstones is significant. This is a U-shaped vertical burrow which tends to move up or down dependent on the addition or subtraction of sediment from above the burrow. Because of this curious habit, Roland Goldring named the form from similar Devonian sediments in North Devon *Diplocraterion yoyo!* (1962). It is an excellent indicator of intertidal conditions. There are also several molluscs which are similarly instructive. I am tempted to quote the old music-hall tongue-twister song:

She sells sea-shells on the sea-shore
And the shells she sells are sea-shells I'm sure,
For if she sells sea-shells on the sea-shore,
Then I'm sure she sells sea-shore shells.

That song may date me a little, but it illustrates my point that one can recognize a sea-shore by its shells and by other organisms as well.

Probably the most traumatic thing that can happen to benthic marine organisms is for the tide to go out. They are then in immediate danger of desiccation. The most obvious tactic for avoiding this is for them to burrow rapidly into the welcoming damp sand. There are then the further dangers of resurrection by erosion when they must burrow farther down, or of being buried alive too deeply by further deposition and they must move upwards in the sediment. Hence the yo-yo effect.

Diplocraterion is also characteristic of the intertidal zone of the Miocene Marada Formation of Libya (Figure 6.3). When Peigi Wallace and I were working on the trace fossils of the Upper Jurassic in the Boulonnais (northern France) we noted that the maximum depth for the *Diplocraterion* burrows in the intertidal levels was about 30 to 35 cm. This struck us rather forcibly as we had just read Rhoads' papers (1966, 1967) giving similar figures for modern intertidal burrowers on the coast of Massachusetts. What is more, we noted a whole series of U-burrows ranging from the horizontal burrows known as *Rhizocorallium* in deeper and/or quieter waters through forms increasingly oblique to the bedding to the completely vertical *Diplocraterion*. We presumed that this represented a story of shallowing upwards to an intertidal situation.

There seems to be a considerable range, however, in the maximum depth of intertidal burrowing. On the basis of too few observations I have noted much shallower burrows in Palaeozoic sediment than in those of the Mesozoic and later. Obviously this may be related to several different factors. It may be a matter of an impermeable or impenetrable layer down below; it may be a matter of height up a sandy beach, though the few bait-diggers I have spoken to do not confirm such a correlation; most exciting perhaps is the possibility that the depth of burrowing may be related to tidal range. I have almost convinced myself that burrowing is shallower along the margins of seas, such as the Mediterranean, with limited tidal range and greatest in areas of considerable tidal range, as outside my study window at Swansea. In that connection it may be noted that the similarity of the Boulonnais Late Jurassic burrows to the modern ones on the other side of the Atlantic may indicate that there was already a fully developed ocean at that time, as I have suggested elsewhere (Ager 1975a). I have certainly seen shallower *Diplocraterion*-type burrows in the above mentioned Miocene Marada Formation of eastern Libya (Figure 6.3). My then research student Harry Doust described (1968), but unfortunately did not publish, a whole series of

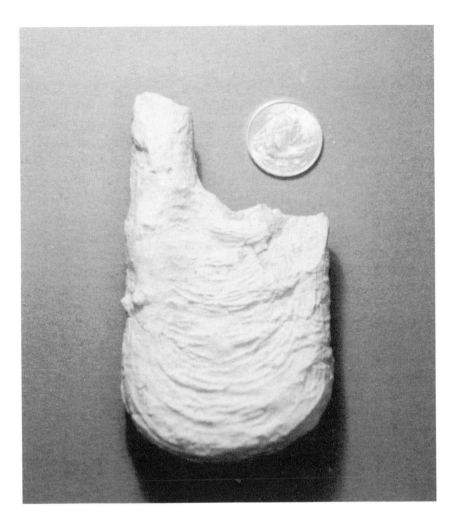

Figure 6.3 *Diplocraterion*, a vertical U-burrow extracted from the Marada Formation (Miocene) of eastern Libya. The scratches on the burrow suggest that it was made by a crustacean. The coin is an old half-penny, one inch (23 mm) across. Photo DVA.

trace fossil assemblages in the Miocene Marada Formation of Libya in which the intertidal facies with *Diplocraterion* came between the deeper marine assemblages on the root structures of the terrestrial facies. Again I wondered here if the shallower burrowing was related to the already enclosed Mediterranean, immediately before the complete drying out of that sea in the 'Messinian Salinity Crisis', when the whole region became a desert with the deposition of extensive evaporites. I have also noted that vertical U-shaped burrows are even shallower in the Lower Cretaceous Hunevtsi Formation of Bulgaria, south-west of Sapelovo and might be referred to the ichnogenus *Corophioides* (Ager 1972, p. 261). Palaeogeographical reconstructions of this area at the time (e.g. in Vinogradov

1961) show a narrow sea that was very much enclosed by land masses.

Burrowing bivalves are extremely common in a variety of shoreline deposits and are commonly preserved in their position of life. This applies particularly to the 'myid' bivalves of the Order Phaladomyoida, most of which apparently had similar habits since way back in Palaeozoic times. Especially common are genera such as *Pleuromya*, with the elongated form and posterior gape characteristic of such burrowers. Particularly fine examples are seen in the Quaternary Yabu Formation of the Bozo Peninsula on the east side of Tokyo Bay in central Japan. These include *Panopea japonica* (Figure 6.4) and *Mactra veneriformis*, which only live today in intertidal sand-flats in the inner part of Tokyo Bay.

Figure 6.4   *Panopea japonica*, a burrowing bivalve in life position, Yabu Formation (Quaternary), Bozo Peninsula, Japan. Photo DVA.

Bivalves of this kind are very liable to be buried too deeply and burrow upwards to survive, forming what are known as 'escape burrows'. Broadhurst & Loring (1970) used them to demonstrate rapid and episodic sedimentation in the Upper Carboniferous (Pennsylvanian) of Lancashire, in the north of England. I have seen them very abundantly in part of the Numidian Sandstone (Upper Eocene) of Tabarka Island near Tunis.

Other bivalves do not burrow so deeply, notably *Pinna*, the triangular shell which is only partly buried in the sediment in life, so that when they are found in life position as fossils, as in the lowermost Jurassic of Southerndown in South Wales they suggest very rapid marginal sedimentation (Ager 1986b). Similar examples occur in the Upper Jurassic of the Boulonnais and East Greenland. At the base of the Lower Eocene London Clay of Whitecliff Bay in the Isle of Wight, more than 90% of the specimens of *Pholadomya margaritacea* and *Panopea intermedia* are still preserved in their burrows (Ager 1963, p. 87); a metre or so higher up is a bed of *Pinna affinis* also in life position. Clearly here coastal sedimentation was very rapid. A few years ago, after an earthquake, the coastline at Napier in New Zealand suddenly rose about 1.5 metres and a new beach appeared covered with the upstanding shells of the bivalve *Atrina*, which is closely related to *Pinna* and is similar in shape. These must have been living just below low tide level.

Of course organisms do not adopt such tactics on the spur of the moment. But there are a whole range of animals which have evolved a burrowing capability through the floods and ebbs of countless tides in the aeons of geological time. Best known perhaps are the segmented worms such as *Arenicola* which is familiar on beaches by the casts of sand from its burrows (which I have noticed seem to be characteristically larger on the shores of the States than in Europe!). Its close relation *Arenicolides* spends so much of its life underground that it has no eyes. These form smooth U-shaped burrows and many similar fossil traces have been called *Arenicolites*. They can be distinguished from the *Diplocraterion* type burrows because the latter have scratches on the sides or *Spreiten* suggestive of crustacean claws. But there are many other organisms which burrow, notably the brachiopod *Lingula* which holds the record for longevity. I saw them burrowing actively in an aquarium in Japan and I have a large specimen of *Lingula shantungensis* from Kyushu preserved in spirit on my desk. The Japanese are renowned for their delicious sea-food meals and I was promised *Lingula* soup, but I did not get it, rather to my relief because, as a brachiopod specialist, it would have seemed like cannibalism.

There were probably many other brachiopods that were adapted to coastal habitats in the past (Ager 1967). These probably included the pentamerids with their beaks heavily loaded with secondary shell material which may have served as an anchor. A well-known British example is *Kirkidium knighti* in the Late Silurian Aymestry Limestone of the Welsh Borderland. In the transgressive Early Silurian, *Pentamerus oblongus* is commonly found in life position and its thick hinge-plates and median septum produce a broad arrow appearance, which give the bed concerned the name 'Government Rock', since arrows like this formerly appeared on British government property such as convicts' suits. Ivanova *et al.* (1962) refigured from Gurbevsk the Devonian pentamerid *Colchidaella* in life position in the Devonian of the Kuznetz Basin. John Ferguson

(1978) showed how spiny productids were adapted to a shallow sandy floor, since they stood up on their spines like ships in dry dock, whilst ordinary non-spinose productids were quickly buried. Peigi Wallace and I have suggested that the very wide, straight hinge-lines of some spiriferids may have served the same purpose (Ager & Wallace 1966b).

Crab burrows are probably more common in the record than is generally supposed. I have seen them in the Miocene Marada Formation of eastern Libya, where they reminded me of the dozens of small crabs that scuttle and rapidly disappear on tropical beaches such as those of Kenya and Tanzania. The problem here is that such beaches are usually composed of loose uncompacted sand which do not readily lithify. One way that the animals had for dealing with this difficulty was to form pellets, cemented with mucilage, to line their burrows. This is seen, for example, in the trace fossil *Ophiomorpha*, which occurs in both horizontal and vertical orientations. Wallace and I suggested (1971) that in the Upper Jurassic of the Boulonnais, the smooth *Thalassinoides* of slightly deeper fine-grained sediments became the 'knobbly' *Ophiomorpha* in intertidal sandy sediments. Somewhat comparable are the pellets made by various species of crabs as they eject the sediment from their burrows. One commonly sees rings of such pellets around their holes, for example around the holes of the tree-climbing crabs of Tahiti, which patronize the palm trees along the sea-shore. It is tempting to use this interpretation for many sandy sediments entirely composed of small, sphaeroidal nodules. Thus in the Lower Cretaceous of southern Tunisia and in the Boulonnais there are whole beds of such pelloidal accumulations in a coastal environment, but one must be sure that they are not just a diagenetic effect of poikilitic recrystallization.

King (1965) described interweaving limulid trails in the Devonian Bude Sandstone in south-west England and interpreted them as mating trails (this has been called 'palaeopornography'). At the top of the Upper Triassic (Norian) continental Mercia Mudstone of Westbury on the River Severn in Gloucestershire there are abundant arcuate markings in a ripple-marked and rain-pitted sandstone. This is part of the transgressive Westbury Beds (Rhaetian) and fits in well with the concept of an ephemeral sandy coastline (Ager & Edwards 1986). I have interpreted these markings as the impressions of the head-shields of the 'king crab' or xiphosurid *Limulus* (Figure 6.5). This lives today in great abundance, for example, in the broad embayment of Chesapeake Bay in Maryland, where they come ashore for mating. They are also common in Long Island Sound, whilst on the other side of the Pacific they are found along the coasts of the Sea of Japan, between Japan and the Chinese mainland and in the Gulf of Tonking in Vietnam. The large female crawls up the beach to deposit her eggs, which are fertilized by the following male. They do this in the early summer when the moon is full and the tides are highest. I like to think of these creatures pursuing their natural

Figure 6.5   Impressions left by the xiphosurid *Limulus* or 'king crab' in the Upper Triassic (Rhaetian) of Westbury-on-Severn in the west of England. The specimen by the chisel (which is 16 cm long) shows the telson or caudal spine. It is thought that they crawled up on to the beach for mating. Photo DVA.

instincts on this English Triassic sea-shore blanched in the moonlight of a warm summer night. It is not often that one can be so precise in one's interpretation of the past!

Also common in the Miocene Marada Formation of Libya are the curious flat echinoids with holes, the scutellids or sand-dollars. These are good indicators of sandy shorelines and though we could not find one actually in a burrow, Harry Doust showed me wide flat burrows in which they would certainly fit.

The effect of burrowing organisms on shallow water sediments is and was immense and as a broad generalization it may be said that bioturbation tends to become more intense as one approaches the contemporary shoreline. Another important effect that the proximity of land has on marine faunas is that of freshwater run-off and therefore of decreased salinity. This is seen particularly in estimates of diversity, which is characteristically low near shore. It then increases to an off-shore peak before decreasing again with increasing depth. This may well be the explanation

I noted in the Cerro Gordo Member of the Hackberry Formation (Upper Devonian) in Iowa (Ager 1963, pp. 239–42). In the Upper Jurassic of southern England I have suggested (1975a) that the fall in diversity (and successive disappearance of significant groups) is directly related to a change in salinity, in this case upwards to the evaporites at the top of the System.

When I started work on the Jurassic of the Moroccan High Atlas, first with research students Chris Burgess and Chris Lee and later with Tony Adams, I was surprised at the amount of the calcareous succession which could be attributed to an intertidal and supratidal environment. The Chrises found this in the Lower Jurassic of the central High Atlas, where great scarps represent this facies (Figure 8.4). They are all thin-bedded limestones and evidently represent episodic sedimentation. In the western High Atlas, similar beds in the Upper Jurassic were evidently deposited behind small coral reefs and were lagoonal in character. They contained tall nerineid gastropods, which I associate particularly with low salinities. They also displayed large megalodontid type bivalves, which were supposedly extinct before this time. Again they were episodic in nature with repeated layers of what I interpreted as algal mats. These had been repeatedly ripped up and redeposited in graded beds, which I interpreted as the effect of repeated storms hitting the coastline (Ager 1974). I will discuss this further in Chapter 9.

Conclusions

From all that has been said above, it is obvious that ancient shorelines are rarely preserved, partly by reason of their very nature, but even when they are preserved they are only seen as fleeting facies that come and go. Like everything else in this book they come and go in an episodic manner.

7
Ancient platform deposits

For many years I have pointed out how, at certain times in earth history, carbonate deposition was remarkably widespread around the world (Ager 1973, 1975a, 1981a, 1981b). Thus the Mid Silurian Niagaran limestone of the North American Great Lakes region (over which tumble the Niagara Falls) is the near equivalent of the Wenlockian limestone of Britain (forming the feature known as Wenlock Edge) and what were formerly called the Gotlandian limestones of Sweden, seen in the cliffs of the Isle of Gotland in the middle of the Baltic Sea.

In the Frasnian Stage of the early part of the Late Devonian, limestones of this age form the cavernous reef limestone of the Fairholme Formation of the Canadian Rockies, as in the striking escarpment of 'Chinaman's Leap' near Banff (Ager 1981a, Plate 1.11) and the chief source of oil under the prairies of Alberta (Figure 7.1). The same stage is seen in a minor way in the limestones of Plymouth and Torquay in south-west England, the classic reefs of Belgium, the hugely fossiliferous Calcaire de Ferques in the Boulonnais, northern France, the beautiful karst country of Moravia in central Czechoslovakia and goes on all the way to Kashmir in India and the famous reefs of the Windyana Gorge in the Kimberley Mountains of Western Australia (Ager 1981a, Plate 1.12).

In the Lower Carboniferous (Dinantian or Mississippian), the Redwall Limestone, which forms the steepest cliff in the Grand Canyon is the equivalent of the escarpment made by the Rundle Limestone in the Canadian Rockies (Ager 1981a, Plate 1.10) and the Lisburne Limestone of Alaska. The so-called 'Carboniferous Limestone' of Britain used to be called the 'Mountain Limestone' because it forms so many of the cave-riddled uplands of England and Wales. The latter name was also used at first (presumably by geologists with a European-oriented education), before a plethora of formation names confused the issue, for the widespread carbonates of the same age in the American Mid-West, such as those which accommodate the Mammoth Cave of Kentucky. In Belgium, the 'type section' at Dinant forms a massive cliff, on which stands the citadel above the River Meuse (Figure 7.2).

The thick carbonates of the Late Palaeozoic ended with the Permian

Figure 7.1   Block of the Fairholme Formation of Late Devonian (Frasnian) age, outside the Geological Survey in Calgary, Canada. The cavernous reef rock is the main source of Alberta's wealth in the form of oil. This is the only monument to a rock that I know of anywhere in the world. Photo DVA.

Figure 7.2   The cliff of Dinantian (Lower Carboniferous) limestone at Dinant in Belgium above the River Meuse. Photo DVA.

glaciation. They were not seen again until Mid Triassic times (locally as the Muschelkalk), and the thick limestones and dolomites of the Upper Triassic. I have suggested in the part of the stratigraphical column I know best, i.e. the Mesozoic (Ager 1981b) that there were climatic optima in the Norian (Late Triassic), Bajocian–Bathonian (Mid Jurassic), Tithonian (end Jurassic), Barremian–Aptian (the so-called 'Urgonian' of late Early Cretaceous) and Campanian (Late Cretaceous). In this I followed Fischer's suggestion (1979) of Mesozoic cycles of the order of 32 million years. In my interpretation, these were times of widespread carbonate deposition, for example the Late Triassic Haupt Dolomit of the Austrian Alps.

The Mid Jurassic 'Inferior' and 'Great' Oolites form the glorious escarpment of the English Cotswolds and these seem to continue all the way down to the French Jura; they turn up again in Kenya as the Kambe Limestone forming an escarpment, for example south-west of Kilifi (between Mombasa and Malindi), and are again often oolitic.

Oolites and oncolites are a special case, since we know at the present day that they only form in very shallow, agitated water close to the shoreline. Yet oolitic limestones can be as extensive as those mentioned above. Also in the Cotswolds there is a thick bed of algal oncolites known as the 'Pea Grit' (though the oncolites are more like baked beans than peas). This extends all along the Cotswold escarpment. Another oncolite forms an excellent marker for mapping in the Kimmeridgian of the French Jura (Ager & Evamy 1963).

Late Oxfordian limestones are widespread in the Upper Jurassic of Europe and north-west Africa. Coral reef limestones of this age extend from England across France and Germany and sponge reefs are found from Portugal to the startling pinnacles formed by them near Ogrodzieniec in Poland (Figure 7.3). In the High Atlas mountains of Morocco, coral reefs of this age provide a little oil and a similar rock is the Zuloaga Limestone of Mexico which seems to be of the same age as an important source of oil in the Yucatan peninsula.

Figure 7.3   Peaks produced by the weathering of sponge reefs in the Upper Jurassic of Ogrodzieniec, Poland. Photo DVA.

Tithonian limestones, at the very top of the Jurassic, provide the Portland Limestone in England for ecclesiastical and bureaucratic buildings. Limestones of the same age cap many of the boxfolds in the Jura Mountains of France, for example in the mountain above our delightful (and gastronomic) 'headquarters' at Virieu-le-Grand (Figure 7.4) and farther south the narrow gorge through which the Rhône roars at Yennes. They continue into the Alps (albeit with different organic constituents), for example in the 'Porte de France' at Grenoble. The shallow water limestones in England, with their giant ammonites and bivalves become coral reefs in the Jura and *Calpionella* limestones in the Alps. They form the striking frontal escarpment of the Carpathians in Czechoslovakia (Figure 7.5) including the famous Stramberk Limestone, which again has coral reefs, but also a remarkable brachiopod fauna that is not known elsewhere (except very locally in neighbouring Poland). So whatever the faunal constituents, we always find limestones at this level.

Figure 7.4   Folded Upper Jurassic rocks, capped by Tithonian limestone, Grande Montagne de Virieu, Virieu-le-Grand, southern French Jura. This is one limb of a typical Jura boxfold. Photo DVA.

In the Cretaceous, the great limestone mass of Glandasse dominates Hannibal's route through the Alps. This is an 'Urgonian' limestone of Barremian to Aptian age. It goes down into south-east Spain and also caps anticlinal ridges in the Jura such as the Montagne de La Balme near Bellegarde. From there it goes all round the Alps and Carpathians to Georgia in the southern USSR. I illustrated several of these in a previous book (1981a, Plates. 1.1 to 1.4). I was surprised and delighted to find a very similar rock, of similar age as the Cupido Limestone of north-east Mexico, which forms strong features such as the escarpment above Los Lirios (Figure 7.6). The Late Cretaceous limestones are the most famous of all, extending from the chalk of the 'White Cliffs of Dover' west to Ireland and to the Mississippi Embayment of the USA, with deposits such as the Austin Chalk of Texas. Eastwards it extends to Bulgaria, the southern USSR and Turkey (Ager 1958). Most remarkable of all, it then appears in Western Australia as the Gingin Chalk with similar lithology, concretions and fossils. This ranged in age from Cenomanian to Maastrichtian, but the climatic optimum appears to have been in the Campanian (Ager 1981b).

In the Palaeogene there are the Mid Eocene nummulitic limestones on which stand, for example, the beautiful village of Ager (of which I

Figure 7.5   Tithonian limestone in the front escarpment of the Carparthians, Pavlov Hills, Czechoslovakia. Photo DVA.

Figure 7.6   Mid Cretaceous (cf 'Urgonian') limestone forming a ridge near Los Lirios, north-east Mexico. Photo DVA.

am not surprisingly fond), in north Spain, just south of the Pyrenees, the monastery of Zirc in Hungary (where the monks used to make cheese, but is now a cheese factory), the entrance to the High Tatra national park near Zakopane in southern Poland and in the pyramids of Egypt. In the Neogene the climax of carbonate deposition seems to have been in the Miocene, as in the golden cliffs of Malta, the building stone of Vienna and the escarpment of the Marada Formation of eastern Libya. I write only of the rocks I have seen personally (except sadly those in Australia, though I am still hoping). I am sure that all my examples could be extended a lot farther. One could go on almost indefinitely with similar examples.

There is growing evidence from isotope and other data that these episodes were times of climatic optima and that carbonate deposition ended with declining average temperatures. Perhaps the best example is that at the end of the Cretaceous, when a deteriorating climate is more likely to have caused the extinction of the dinosaurs and all the rest than any intruder from outer space. Declining temperatures gave the opportunity to warm-blooded mammals from the north to take over. However, dinosaurs are rare and like all large animals (such as the elephants and whales today) must have been very vulnerable. More important was the sudden slump of marine microfossils such as the coccoliths, which caused the collapse of the whole marine ecosystem. I like Hallam's suggestion (1984) that the extinctions caused the iridium rather than the other way round. In other words, the sudden disappearance of the main rock-building microfossils caused a break in deposition and the accumulation of iridium-bearing meteoritic dust such as is found on ocean floors today. But this is a matter for Chapter 13.

At the present day we are sadly lacking in carbonate deposition, especially in shallow water environments, apart from very local examples such as the over-worked Bahamas Banks in the Caribbean and Shark's Bay in Western Australia. There is also some forming the coral rock down the east coast of Africa and on the west side of the 'Persian' Gulf, but these are nothing compared with the vast extent of shallow water carbonates at particular moments in the geological past.

There is plenty of evidence that the world climate today is cooler than it has been through most of geological time. This is shown by the relative narrowness of the present tropical belt. It is most clearly demonstrated by the evidence of plants, in particular the presence or absence of growth rings indicating the existence of seasons. Thus Creber and Chaloner (1985) concluded from plant evidence that there was a much warmer global climate in the Mesozoic and Early Tertiary with a broad zone of non-seasonal, i.e. tropical, climate stretching from palaeolatitudes 32°N to 32°S. One may note also the presence of plants such as the Permian *Glossopteris* in Antarctica, where such plants could not possibly live today. Even in Permian times that continent was probably at a very high latitude (see for example Scotese *et al.* 1979). An even wider tropics is suggested by the Eocene floras of the Isle of Sheppey and the Hampshire Basin in England (Ager 1963, pp. 192–5). Chaloner & Creber (1990) went on to suggest that the frequency of stomata in fossil leaves was a reflection of changes in the partial pressure of carbon dioxide in the atmosphere as had been proposed by Woodward (1987) for Recent forms.

Similarly scleractinian coral reefs in the Jurassic grew as far north as England and the Queen Charlotte Islands off the west coast of Canada. Though one obviously should beware of possible changes in habit of major groups of organisms with time, there is no reason to doubt that this particular group had similar physical requirements in Jurassic times to those it has today.

Drost-Hansen & Neill pointed out (1955) that a sharp change in the chemical properties of water occurs at 15°C, which is roughly the winter isotherm at the boundary between the tropical and sub-tropical zones. This marks the sharpest break in marine faunas at the present day (Ekman 1953). Stehli (1957) used this evidence to indicate the limits of the Permian tropical belt, which corresponded roughly with the present 55°N latitude. (He was not in favour of polar wandering so far as Permian faunas were concerned). I suggested (Ager 1963) that this also corresponded with other fossil distributions in the Palaeozoic and Mesozoic. I further suggested (*op cit.*, p. 156) that with the higher global temperatures in the past there was formerly a 'hypertropical belt', with higher temperatures than those we know today. The disappearance of that belt with a deteriorating climate might explain the extinction of such equatorially distributed groups as the archaeocyathids in the Cambrian and the rudistid bivalves in the Cretaceous. It may also correspond with the limits of the main belts of carbonate deposition, for example in the Jurassic (Ager 1975a, pp. 1–11 *et seq.* and Figure 5).

It is therefore my main suggestion that the times of widespread carbonate deposition in the past may be related to the availability of carbon dioxide in the atmosphere. Labeyrie & Jeandel (1990) have emphasized that it takes several years for an equilibrium to be established in carbon dioxide between surface water and the atmosphere and up to hundreds of years for exchange with deeper waters, but of course this is no time at all, geologically speaking.

The Gaia hypothesis of Lovelock (1979, 1983, 1989) has been much criticized, for example by Postgate (1988). I have heard it called a religion rather than a scientific theory, but to me it makes a lot of sense. As I understand it, the Earth is thought to be a self regulatory system, like a gigantic organism. Lovelock sees the evolution of the rocks, atmosphere, oceans and organisms as a single tightly-coupled process. He compared the Earth with a tree in which 90% or more of the tree is dead with only a thin skin of living tissue on the outside (1989, p. 749). In sedimentological terms this is where things happen and this is where limestones are deposited if other conditions are right.

I think of the hydrosphere as acting as a buffer zone between the atmosphere and the lithosphere. Thus when carbon dioxide builds up in the atmosphere, it may be taken out in carbonate deposition.

My former colleague Tony Ramsay (1974) showed that fluctuations in the rate of solution of calcium carbonate in the oceans and in the calcium carbonate compensation depth have occurred since Jurassic times, probably at intervals of 250 000 years or less. He attributed this to changes in plankton productivity, particularly in high latitudes, resulting from global variations in climate through time. Warm periods were characterized by increased plankton productivity and the deposition of carbonates in high latitudes, whilst cold periods were characterized by a decrease in plankton productivity and the distribution of carbonates in mid latitudes. If one looks at the Upper Cretaceous, one sees carbonates widespread in the higher latitudes of Europe and North America, but earlier in the South Atlantic there were thick carbonaceous deposits, but not carbonates. This relationship between plankton productivity and temperature fits in with my remarks earlier about the events at the end of the Cretaceous.

Now I must turn to the question of volcanicity, for the end of the Cretaceous was also the time of the great out-pourings of the 'Deccan

Traps' in the Indian sub-continent. Walker (1977) and Holland (1984) attributed the level of carbon dioxide in the atmosphere mainly to volcanic activity and the calcium to the weathering of calcium-bearing rocks, presumably in the first place chiefly the anorthite in plagioclase felspars and secondarily to pre-existing limestones and dolomites. It has been said that carbonate regimes tend to be self-perpetuating and that is certainly the impression one gets, for example in the Mesozoic rocks of southern Europe.

The relationship of volcanicity to the composition of the atmosphere has often been discussed (e.g. by my friend Jan Brunn 1983), but apart from examples such as the Deccan Traps, it is difficult to provide firm evidence of a correlation with carbonate deposition on a world-wide scale. I have previously emphasized that much pollution is 'only natural' (Ager 1985b). If Walker and Holland are right and volcanicity is the main producer of carbon dioxide in the atmosphere, then the present limited distribution of carbonate deposits may be related to the present comparatively low level of volcanicity. However, on the whole the geological evidence seems to point to temperature rather than volcanicity as being the main control of carbonate deposition, though this in turn may be related to volcanic activity.

I suggest that the times of widespread carbonate deposition in the past were related to the availability of carbon dioxide in the atmosphere. This may have resulted from a slow build-up of this gas (and perhaps methane) such as we see going on today, coupled with the increase in global temperatures due to the 'green-house effect' about which we now hear so much. We may therefore expect, in the near geological future, a rise in sea-level and another spread of carbonate deposition on the continents, as happened so often in the past. I do not necessarily blame all this on mankind.

Even if my hypothesis is not acceptable, there can be little doubt that widespread carbonate deposition coincided with periods when global temperatures were higher than usual and there is another such period coming, but I shall not live long enough to have my ideas disproved!

8

Off-shore deposits, ancient and modern

In this chapter I propose to include both ancient and modern off-shore deposits, largely because we are forced so much to hypothesize about the one on the basis of what we know about the other, in both directions.

Neocatastrophism really started with off-shore deposits, at least it did so far as I was concerned. When I was a student, we were still learning about deposition being a gentle continuous process, with sedimentation keeping pace with subsidence. Then along came turbidity currents with a whoosh and we realized that it could all have happened on that Tuesday afternoon which I keep on about. We were now allowed to have a trench which was filled suddenly rather than gradually.

Though now so much a part of geological lore, turbidity currents were first really recognized by the progressive cutting of submarine cables off Newfoundland on the afternoon and evening of 18 November 1929 (I am not sure if that was a Tuesday afternoon, but it hardly seems worth the trouble of finding out). The telephone connections between North America and Europe were cut, one after another, and it would have been disastrous for anyone wishing to make an urgent call at that time, as my parents-in-law might have done if they had close relations in North America, waiting for news; but I am forbidden to mention that! It became apparent that a current of dense sediment laden water had probably swept down from the Grand Banks, triggered by an earthquake. Such turbidity currents had been recognized long before, in 1840, by a Swiss engineer, Forel, who noted how the cold, muddy waters of the Rhône suddenly disappear as they enter the warmer water of Lake Geneva (see Putnam 1964, p. 379). One can see this happening almost anywhere in the world where muddy rivers enter lakes, especially when the rivers have come from colder mountain areas or areas where there is rapid erosion of the bed-rock. Thus the muddy waters of the Colorado River are rapidly filling Lake Mead, behind Boulder Dam, with the sediment they have picked up in their hectic rush through the Grand Canyon. Bored geology students (not mine of course) often make their own turbidity currents by stamping on the muddy edges of pools.

The erosive power of turbidity currents may be quite phenomenal and is linked to the problem of submarine canyons. Starting in Monterey

Bay, south of San Francisco in California, a submarine canyon has cut
down through a granite of the Coast Range and is comparable in size
and cross-section to the Grand Canyon itself (Figure 8.1). I took a
photograph of the water above the canyon, but this is hardly worth
reproducing here. There has been much argument as to whether such
canyons were formed actually below the sea or if they were sub-aerially
eroded at a time of uplift and/or low sea-level. All that concerns me here
is the force and duration of the process.

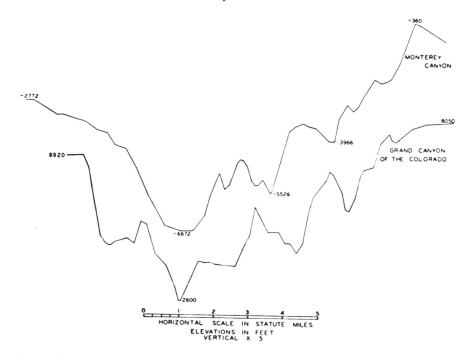

Figure 8.1 Sections across Monterey Submarine Canyon (above) and the Grand
Canyon (below) to show their comparable dimensions. From an American source I have
forgotten. I hope they will excuse me!

The sedimentary product of the turbidity current is, of course, the
'turbidite', though I have often argued against that term on the philo-
sophical principle that one should not use a name which presumes an
unprovable origin. A descriptive name is much more satisfactory, such
as the old-fashioned word *greywacke* which was much used in Britain in
the past, though it has always meant different things to different people.
The outstanding characteristic of such deposits is that each bed is graded,
becoming finer-grained upwards. The late Professor Gouvernet of Mar-
seille showed me modern examples of such 'turbidites' obtained in cores
from the floor of the Mediterranean off that fascinating city. I suppose
I had better give in to the majority and call them 'turbidites', though
personally I prefer the term 'flysch' for the distinctive alternating beds of

harder and softer layers produced by the grading. I should note, however, that Rudolf Trümpy has half-seriously objected (personal communication) that this was a local Swiss term, not intended for export.

The best example I know of flysch is on the north Spanish coast, near Zarauz, where my wife and I once spent a delightful holiday (at least she did, I looked at the rocks). Here a thick formation of flysch spans the Cretaceous–Tertiary boundary and is particularly photogenic at Zumaya (Figure 8.2). This was clearly an off-shore deposit produced by rushes of turbid water which settled into graded beds. Episodicity was the vital factor, as in so much of my story. Such sand/clay flysch is seen all over the world in off-shore deposits, for example in Hokkaido, Japan (Figure 8.3). Farther west along the north Spanish coast there are calcareous turbidites, which are also widespread when the sand available is calcarenite.

In the western High Atlas of Morocco, massive limestones with coral and algal reefs are developed in shallow water environments, but in the central High Atlas my former research students Chris Lee and Chris Burgess showed me thin-bedded limestones in intertidal and supratidal deposits (Figure 8.4). These pass into deeper water micrites which were nevertheless clearly graded, as can be seen from the differential weathering of the top and bottom of each bed (Figure 8.5). These may be called 'distal turbidites'.

Locally the flysch or other thin-bedded off-shore deposits are highly contorted due to the later mountain building movements, which were commonly inflicted on these deposits in what we used to be allowed to call geosynclines. Now we should probably talk of trenches on continental margins. Such contortions are well seen in the flysch of the north Spanish coast, but only in a frustrating manner in sections on the motorway (*autopista*) along the coast west of San Sebastian, where the geologist is forbidden to stop. My only reservation here is that one also finds graded beds in shallow water storm deposits, which I have equally wrongly called 'tempestites' (Chapter 9 and Ager 1974, 1981a). I have also seen these in Morocco, Spain, Portugal, Poland and Japan.

Flysch has its own distinctive suite of trace fossils, notably the so-called 'grazing trails' or *Pascichnia*, which I prefer to call 'browsing trails', since there is not much grass at the bottom of the deep sea! The classic work relating trace fossil assemblages to depth was that of Seilacher (1958) and there has been much literature since, but the flysch assemblage (as at Zumaya) is the most distinctive of all.

Long before we learned the word 'turbidite', geologists such as Sir Edward Bailey were getting excited at the distinction between the cross-bedding of shallow water deposits and the graded bedding of off-shore or geosynclinal deposits, as in the Lower Palaeozoic sediments of North Wales. It was in the famous *Macigno* Formation of the northern Apennines in Italy that Kuenen and Migliorini (1950) first recognized turbid-

Figure 8.2   Cretaceous/Tertiary flysch near Zumaya on the north Spanish coast, with a party of Swansea second year undergraduates being instructed by Peter Styles (beard) and Kenny Ladipo (research student from Nigeria). Photo DVA.

Figure 8.3   Late Cretaceous (Coniacian) flysch, Ikushumbetsu, Hokkaido, Japan, with the author as scale. Photo Prof. Makato Kato of Sapporo.

Figure 8.4   Early Jurassic (Pliensbachian) supratidal and intertidal limestones, Jebel Timeraratine, central High Atlas, Morocco, with research students Chris Lee and Chris Burgess and their wives to be on mules. Photo DVA (on first mule).

Figure 8.5   Lower Jurassic (Pliensbachian) deep basinal micrites, Bin el Ouidane, central High Atlas, Morocco. The projecting lower layer of each bed indicates that the beds are graded. Hammer (not mine) probably measures 29 cm. Photo Dr C. W. Lee.

ity currents as a cause of graded bedding. Kuenen did this by experimentation and Migliorini by field observation. The *Macigno* is really impressive, with graded beds up to more than 20 m in thickness. The formation as a whole reaches more than 3000 m and was all deposited in Late Oligocene to earliest Miocene times (Ager 1955b and 1980, p. 482). So clearly deposition was very rapid, with turbidity currents sweeping down from a land-mass to the west of the present Italian western coastline, of which Napoleon's Elba is perhaps the last remnant. In other words the whole formation consists of a series of sudden Tuesday afternoon deposits.

The 3000 m of the *Macigno*, representing such a small part of Tertiary times, may be contrasted with the earlier beds in the northern Apennines where everything from the Late Jurassic to the Early Miocene may be packed into no more than 100 m of sediment. So again clearly something happened very suddenly and with the drama of Grand Opera in the land of Puccini and Verdi.

Though I am treating turbidites as off-shore deposits, if one goes beyond the range of the turbidity currents, then one comes to sediments dominated by micro-plankton. Therefore other characteristic off-shore deposits of the great trenches are the siliceous radiolarites, as beautifully displayed in the northern Pindus mountains of Greece and often contorted like the flysch mentioned above (Figure 8.6). Even fairly 'normal'

Off-shore deposits, ancient and modern   109

Figure 8.6   Sharply folded Mesozoic radiolarites and thin limestones, somewhere in
the Pindus Mountains of northern Greece (I was lost at the time). Photo DVA.

sediments such as the Mesozoic shallow water limestones with coral reefs
of the Jura in south-east France, become thin-bedded and vulnerable to
intense folding as they approach the Alps and pass into a deeper water
Calpionella facies. This can be well seen in the Val du Fier on the eastern
edge of the Jura (Figure 8.7).

There used to be much discussion about the depth of the sea in which
the Upper Cretaceous Chalk was deposited. At one time it was compared
with the *Globigerina* ooze of modern oceans, but then it was thought
to be not so deep, partly on account of the abundant large benthonic
invertebrates it contains, notably bivalves and echinoids. Now we know
it to be largely composed of planktonic coccoliths, which live at certain
levels in oceans of almost any depth. The depth inhabited by the living
organism has, of course, no connection with the depth at which its dead
remains are found. I would simply suggest that the Chalk was laid down
beyond the reach of land-derived sediment and so is dominated by coccoliths and planktonic foraminifers which settled upon the sea-floor
beneath.

The radiolarians, coccoliths and tintinnids such as *Calpionella* were
and are all planktonic organisms, which sank to the sea-floor in vast
numbers on death. Today we have the *Globigerina* oozes which characterize the deep ocean floor. Such deep sea deposits today are usually

Figure 8.7   Contortions in thin-bedded Mesozoic limestones and shales, near the entrance to the Val du Fier, on the eastern side of the Jura mountains, south-east France. Photo DVA.

assumed to be a continuous record of a 'gentle rain from heaven' type sedimentation (see Ager 1981a, p. 58). It was here, if anywhere, we should find gradualistic evolution as discussed in Chapter 10. In fact, as my friend Fred Banner has told me, this seems to be more true for groups such as the planktonic foraminifers than for benthonic groups. Nevertheless, deep sea research has proved major gaps even in the record of oceanic sediments as everywhere else (Ramsay 1977).

Also from the Apennines came the idea of *cunei composti* or composite wedges of tectonic origin and submarine landslips formed under the force of gravity (Migliorini 1949 & 1952). The basic idea (summarized in Ager 1955b) was that the mountain range formed as a series of successive ridges, each consisting of a set of complex wedges of rock pushed upwards from the west. These were followed by great submarine landslips (though another word is needed here, since they were under the sea). In these, huge masses of rock up to the size of mountains, slid down into deeper water forming the great jumble of rocks known as the *argille scagliose* or scaly clays. Similar complex jumbles have been recognized in many other parts of the world such as Taiwan, Timor, Morocco and Turkey (Ager 1958). Slumping is often seen between flat lying beds as in the north Spanish flysch already discussed (Figure 8.8).

Figure 8.8 Slump in Eocene flysch, Oria estuary, north Spanish coast. Photo DVA.

I saw the same effect in the snow on the roofs of buildings when I was trapped in wintry Kashmir!

Submarine faulting or 'landslipping' had been recognized long before by Bailey & Weir (1932) in the fascinating narrow strip of Jurassic rocks in Sutherland, on the east coast of northern Scotland. Here in Late Jurassic times, great boulders of Devonian 'Old Red Sandstone' repeatedly tumbled down into Kimmeridgian deep water deposits (Figure 8.9). Bailey & Weir blamed this on episodic faulting, triggered by earthquakes and they called the deposit a 'natural seismograph'. Previously the largest of these fallen blocks had been called 'the fallen stack of Helmsdale', but it would have had to be a sea-stack of incredible height. There are associated slumps and clastic dykes. Years ago I strung out groups of undergraduates along this coast, with the task of confirming or denying the hypothesis on the basis of the evidence in their particular sections. I think they were convinced!

This is a case where the past may be said to have been used to interpret the present. Lewis (1971) recorded slumping on the continental slope off New Zealand inclined at only 1°–4°. He found four slumps of sediment up to 50 m thick, probably caused by major earthquakes. Simm and my brief successor at Swansea, Rob Kidd (1984) described modern submarine debris flows detected by Long-Range Side-Scan Sonar off north-west Africa.

Figure 8.9   Fallen boulder in Late Jurassic (Kimmeridgian) deep water sediments, on beach near Helmsdale, Sutherland, north-east Scotland. The hammer is 29 cm long. Photo DVA.

Other characteristic rocks of the ocean floor are the sodium-rich pillow lavas or spilites which are erupted in the oceans up the opening cracks as the plates separated. These form sphaeroidal masses with hard crusts due to the rapid cooling action of the sea water. Continuous pressure by the rising lava breaks through the crust to form more pillows. They are seen forming today off Hawaii, where lava flows into the sea. They are also seen in ancient rocks all over the world in the right situations. There are some excellent examples in the younger rocks of the Alps and some in much older rocks (Early Ordovician in fact) at Strumble Head in what geologists still call Pembrokeshire, since it is a meaningful geological unit, but officials now call Dyfed. The best ancient examples I have seen anywhere were in the Eocene of Oamaru, near Dunedin in the South Island of New Zealand. Here the lavas were erupted into a fine-grained white limestone (Figure 8.10) in which they cooled rapidly, slightly cooking the adjacent sediment and its fauna – of bryozoans and brachiopods – though this hardly suggests oceanic depths. Clearly they can form in shallow water as well.

Associated with the oceanic lavas today are 'burgundy coloured' manganese mudstones. But one has to be careful, it is too easy to find what

Figure 8.10   Eocene pillow lavas, Oamaru, north of Dunedin, South Island, New Zealand. Photo DVA.

one hopes and expects to find. A geophysicist I knew in Morocco, who had much experience of modern oceanic research, but little of land geology, hoped to find evidence of an opening Red Sea type ocean along the line of the Atlas mountains. He confidently identified pillow lavas and manganese mudstones in the Triassic. But the pillow lavas were ordinary terrestrial basalts, which had weathered into spheroidal masses, as they usually do in such arid climates, and the manganese mudstones were perfectly ordinary Triassic continental red beds. Not far away were dinosaur footprints and I could not resist asking him if he thought that these were submarine dinosaurs or dinosaurs with very long legs! I only include this as a warning against finding what one expects to find. But I must not criticize, I'm not much good at geophysics!

Again going back to those far-off days when I was a student, we used to be told that modern deep-sea deposits were characterized by meteoritic dust and whales' earbones. I don't know about the latter, except that they are presumably very resistant structures, but the former would solve lots of problems. The theory was that sedimentation in the deep sea was so slow that material as exotic as meteoritic dust would slowly accumulate in significant quantities. Presumably that dust contains a comparatively high proportion of elements such as iridium which is nowadays so fashionable and a long break in sedimentation would allow for the gradual accumulation of that element. The argument is 'how long were the breaks?' Ken Hsu insists (1986) that there was 'No chasm at Gubbio' (where the whole business of iridium anomalies began). Many of us think that the slow accumulation of meteoritic dust might explain the iridium peak which has been made so much of in blaming the extinctions at the end of the Cretaceous on the catastrophic arrival of a major iridium-rich extra-terrestrial body. Again I must say that I like Tony Hallam's (1984) suggestion that it was the extinctions that caused the iridium and not the other way round. In other words, the extinction of so many rock-forming organisms such as coccoliths and foraminifers resulted in a break in sedimentation and therefore a chance for the iridium to accumulate. However, I trespass on the subject matter of Chapter 13.

Another feature of modern off-shore deposits that always interests me is that apart from the large areas where virtually nothing is accumulating at all, where there is sediment, it is usually on the move. Thus the sand banks of the North Sea all seem to be moving and an example close at hand (which I therefore fully believe in) came from the work of my friends Charitha Pattiaratchi & Michael Collins (1987) in the Bristol Channel, on sand-banks just outside Swansea Bay. They showed that the largest of these – the Scarweather Sands – are constantly on the move, with the sand circulating endlessly in a clockwise direction. This is in an area of high tidal range (here about 10 m) and storm waves come in from the Atlantic. They contrast these local banks with the 'dead' sand-banks out in the Celtic Sea, south of Ireland. In both cases no stratigraphy is accumulating for the future. It is one of those all-important gaps.

One interesting feature of modern oceans, only discovered comparatively recently are the 'black smokers', volcanic gases and metal enriched hot brines emerging in isolated places from the floor and giving a home to a highly distinctive fauna adapted to this unusual environment. The only possible ancient example I know is in the Upper Jurassic of the French Alps in the formation known as the *Terres Noires*, described by Christian Gaillard and several others (1985, 1986, 1988). They suggest that hydrothermal springs have led to 'oases' of diverse large life forms in beds that generally only yield small and sparse faunas.

So I may conclude that off-shore sediments and fossils, like everything else in this book, are episodic, local and short-lived phenomena.

9
Sudden storms are short

Hurricanes, typhoons and storms

As Professor Henry Higgins pointed out (in Shaw's *Pygmalion*): 'in Hertford, Hereford and Hampshire, hurricanes hardly ever happen'. On the other hand, Professor Miles Hayes (1967) showed that if one stood at any specific point on the Texas coast for 3000 years, one had a 95% chance of being caught at some time in the central axis of a hurricane (Figure 9.1). In both cases, therefore, hurricanes may be regarded as rare events.

This was brought home to us in England's 'green and pleasant land' (I was in Idaho at the time!) by the events of the terrible night of 15/16 October 1987, when the winds in the south-east reached unheard of speeds and something like 17 million trees were blown down, including six of the ancient oaks of Sevenoaks in Kent which gave the town its name. This was generally thought to be the hurricane Emily, wandering from the normal course of her predecessors. It caused damage in the gardens of both my brother and my brother-in-law north-west of London so certainly reached Hertford and I know that it caused a lot of destruction in Hampshire. I am told that it did not reach Hereford in the English 'Far West', but clearly Professor Higgins was justified in his statement. So what hardly ever happens, does happen if one waits long enough. We began to get a little worried when it happened again on the night of 25/26 January 1990; that even reached us in west Wales and necessitated expensive roof repairs, including damage to the television aerial, so preventing us from escaping the Welsh-speaking Channel 4 for its English-speaking equivalent across the Bristol Channel (most of the aerials in anglicized Swansea point that way). Everyone talked gloomily of global warming and its effects, but there was some consolation in the fact that Daniel Defoe had described a similar great storm back in 1703 which killed some 8000 people in southern England. His 272 page report was entitled: *The Storm: or, a Collection of the Most Remarkable Casualties and Disasters Which happened in the Late Dreadful Tempest Both by Sea and Land*. He concluded 'That the winds are a part of the Works of God by Nature' (we went even further than the Germans in our use of capitals in those days). So how rare is rare?

116   The New Catastrophism

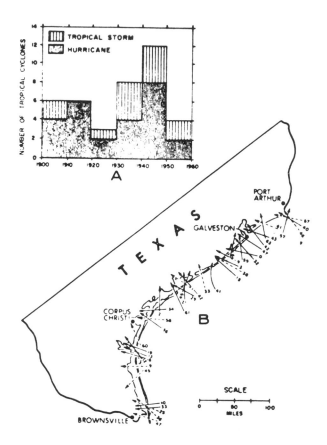

Figure 9.1   Paths of hurricanes or tropical cyclones striking the Texas coast, USA between 1900 and 1960 with a histogram of their frequency. From Hayes (1967) by kind permission of the author and the Bureau of Economic Geology, University of Texas.

In the Pacific the equivalent of hurricanes is the typhoon, as in Joseph Conrad's famous short novel. It was a typhoon and the fire that commonly follows such natural catastrophes, that, in the seventeenth century, destroyed all but the stone facade of Sao Paulo cathedral in Macao, a short trip from Hong Kong across the Pearl River estuary (now a good oil prospect). I was delighted to find that, though a former Portuguese colony, in Macao they drive quite properly on the left.

Hurricanes and typhoons are generally, however, very localized phenomena on the Earth's surface. This was brought to my attention by my research student Tony Adams when I first postulated hurricanes to explain features I saw frequently repeated in the Upper Jurassic rocks of the western High Atlas of Morocco (Ager 1974). These were better illustrated later (in Ager 1981a, plates 4.5 and 4.6). They are also illustrated here (Figure 9.2). We were there at the narrow western end

Figure 9.2   Ripped-up algal mat or 'tempestite' in the Upper Jurassic below Imouzzer-des-Ida-ou-Tanane, western High Atlas, Morocco. The coin is a Moroccan dirham, 2.5 cm across. Photo DVA.

of the Tethys, where wind and wave phenomena, including tsunami (discussed below), might be expected to be concentrated. Adams suggested that we would probably have been outside the hurricane belt in Late Jurassic times and I compromised with the more prosaic term 'storms'.

The features I described in the thin-bedded limestones were repeated 'rip-up' structures. Laminated black bands, which I interpreted as algal mats laid down in lagoons, occur at the bottom of each unit, which average between 30 and 60 cm in thickness. The ripped-up fragments are graded upwards in fine micrite and resemble the graded structures seen in deep-water turbidites. I therefore coined the term 'tempestites' for them, though – as I have said before – both 'turbidites' and 'tempestites' are objectionable terms philosophically, since they presume an origin that cannot be proved. But at least the term is evocative and memorable, which is more than can be said for the 'HCS' jargon (for 'hummocky cross stratification') which has grown up in recent years and is said to be characteristic of storm deposits in siliciclastic sediments, but is notable for its absence in the examples I know. What I would really favour, though I doubt if it will catch on, is 'Tuesday afternoon deposits' which I used for some Lower Jurassic storm deposits near my home

city of Swansea (Ager 1986b). This was merely meant to convey the
idea that these deposits were laid down in a very short period of time
and somehow Tuesday afternoon sounds the silliest afternoon of the
week.

The best section I know for my 'tempestites' in the western High Atlas
is by the road leading to the waterfall below the village of Imouzzer-des-
Ida-ou-Tanane ('children of the waterfall of Tanane'). The waterfall
tumbles over a tufa apron which locally conceals the Upper Jurassic
limestones. It provides a very refreshing shower after a day doing geology
under the hot Moroccan sun! So far as I could judge, the limestones are
all contained within one ammonite zone, laid down in what is usually
estimated as about a million years. There are two or three hundred beds,
each – in my view – representing a single storm or hurricane. That
would work out at one every 3000–5000 years, which is about the same
frequency as the hurricanes documented by Hayes (1967) from the Texas
coast. I would not like to push my luck too far on this one, but at least
it is of the same order of magnitude.

In the presentation of one of Sir Edward Bailey's last papers (1953)
to the Geological Society of London, he commented: 'To see a thing
you have to believe it to be possible'. So it is with tempestites. Once I
had seen them in Morocco, I saw them everywhere. I saw them in the
Carboniferous of northern France and in the Spanish Pyrenees; I saw
them in the Jurassic of southern Portugal and the Polish Carpathians
and I saw them in the Tertiary of Japan. Closer to home there are
laminated sands in the Severn Estuary which have been interpreted as
storm deposits laid down on top of clay with peat and tree stumps
dated as more than 8000 years old. My graduate students and research
associates saw them too, in the Lower Carboniferous of South Wales
(Glanvill 1981, Wu 1982) and in the Middle Jurassic of the English
Cotswolds (Mudge 1978). This was certainly not because of any pressure
from me. Indeed, I deliberately say to my research students: 'I'll think
the better of you if you disagree with me'. I have never been one of those
'Second only to God Almighty' type professors, though obviously they
had to justify their disagreement. Since then there has grown up a great
literature about storm deposits and I have a thick wad of cards about
them in my card index, but it is certainly far from complete. I was
particularly fascinated by Darwin's mention (in his book on coral reefs
1874) of a description of Bermuda in 1837 by a Lt Nelson of the Royal
Engineers (not Horatio who died in 1805, although he had been that
way many times previously). It is said that the whole island consists of
'sand drifted by the winds and agglutinated together . . . there occur in
one place five or six layers of red earth, . . . including stones too heavy
for the wind to have moved. . . . Mr. Nelson attributes the origin of
these several layers, with their embedded stones to violent
catastrophes . . .'. Darwin went on to say that 'further investigation has

generally succeeded in explaining such phenomena by simpler means', though he does not say what. He was always the gradualist. In a brief visit to Bermuda, on the way back from Texas, I was pleased to be able to put myself on the side of Mr Nelson. Hurricanes hit that very pleasant and very British island (it was our first colony) every 20 years or so, causing much damage. I believe, as always, in what I have seen for myself, rather than what I read in the literature. Shorelines are particularly vulnerable when it comes to hurricanes. They only build up over the sea and die out over the land, but great damage can be done in the passage from one to the other. This was shown by Hayes (1967) in his studies of the effects of the hurricanes 'Carla' and 'Cindy' in 1961 and 1963 respectively. (Since then the feminists have objected and such disastrous happenings must not now necessarily be blamed on ladies.) One such 'tempest' can change the whole geography of a coastline, as is well known in the short time we have been studying such things. One wonders how many of the disasters recorded in the Earth's histories, mythologies and religions were due to such natural causes, whether they be the one that Noah survived or the one in which Atlantis perished. Mention of Atlantis turns my thoughts to tsunamis or storm surges, one of which might well fit Plato's description of the drowning of that land 'in a single day and night of misfortune' (Kukal 1984).

Tsunamis

The importance of tsunamis or storm surges has probably been underestimated. The power and frequency of these great waves of water, often triggered off by earthquakes or explosive volcanism, has only been appreciated in those countries where they are only too familiar. When I visited Shizugawa in Hokkaido in 1981, the town had only recently been damaged, with casualties, by a tsunami that resulted from an earthquake in South America and which had taken two days to cross the Pacific. The great Meiji Sanriku tsunami of 1896 killed more than 27 000 people in Japan and there were many other such destructive waves in lands such as Chile, Hawaii and the Philippines. Ridgeway (1984) counted no less than 245 in the whole Pacific between 1900 and 1983, an average of very nearly three a year. Only very recently was the disastrous flood in Bangladesh, sweeping up from the Indian Ocean.

Ambraseys (1962) listed 229 tsunami in the Mediterranean in historic times. It is useful to remember at this point that the Mediterranean, although seemingly an enclosed and protected sea, is not always the peaceful place that is commonly imagined. The worst storm I have ever experienced at sea was in the Mediterranean in 1946, when I was coming home in a troopship from the Far East in charge of a troop-deck of 90 men, 89 of whom were sea-sick. My lone sailor daughter Kitty has experienced worse weather there, at least twice, than when crossing the

Atlantic, the notorious Bay of Biscay and the English Channel. Her small yacht was repeatedly damaged by the supposedly placid Mediterranean.

It may well have been a tsunami, caused by and added to the volcanic explosion of Santorini (Figure 11.1) that destroyed the first European (i.e. Minoan) civilization on Crete in the seventeenth century BC. After years of argument, the explosion and the destruction are now again thought to have coincided (Manning 1989). This is a good candidate for the fabled country of Atlantis and will be further discussed in Chapter 11.

People whose job it is to worry are now concerned about the danger of such surges and about the possible effect of such high and sudden floods on the many nuclear reactors that are spaced around the shores of Britain. A plentiful water supply is essential for these, but not too much at once!

Turning to the ancient record, my latest and definitive 'Tuesday after-noon deposit' is the one in the Lower Jurassic at Ogmore-by-Sea, near my home at Swansea (Ager 1986b). My interpretation has been criticized by the Survey establishment, but my head is unbowed. At this locality there is an eroded surface of Lower Carboniferous limestone, gouged out by wadi-fills of Triassic breccia which came off a desert upland near the end of that period (mentioned in Chapter 4). Then there was an invasion by an Early Jurassic sea, which deposited a conglomerate known as the 'Sutton Stone'. This has usually been interpreted as the basal conglomerate of a diachronous transgressive sea. It has been suggested, with very little fossil evidence, that this conglomerate spans three to five ammonite zones and therefore up to five million years in time. I think it was deposited in a matter of hours or minutes. My reasons for thinking this is that it is not the sort of deposit that one sees on a modern pebble beach, in the very next bay for example, (Figure 9.3). In such deposits each pebble rests on other pebbles. On the coast at least the Sutton Stone is a 'matrix-supported conglomerate' with the pebbles 'floating' in a fine-grained matrix (Figure 9.4). In confectionery terms it is not a tube of small, individual 'Smarties' in contact, it is more like a bar of fruit and nut chocolate. It is also noteworthy that there are no bedding planes within the Sutton Stone in the coast sections. Such matrix-supported conglomerates are characteristic of mass flow deposits and I think in this case that a great wave of water swept in from off-shore carrying fine-grained calcareous silt and ripped off pieces of the wave-cut platform and redeposited them together to form the conglomerate. Perhaps it was a hurricane or a severe tropical storm, perhaps it was a tsunami. Whatever it was I think it happened very quickly.

That brings me back to the title of this chapter. People who know their Shakespeare will remember John of Gaunt saying to the Duke of York in *Richard II*:

Figure 9.3   Modern storm beach of pebbles in contact with each other, top of beach at Southerndown, Mid Glamorgan, South Wales. Photo DVA.

Small showers last long,
But sudden storms are short.

That obviously applies to storm deposits, but it also, in my view, applies to geological processes generally. The 'small showers' lasting long are the day by day processes, in human terms they are the ploughing and sowing, reaping and mowing, the building and decay of castles and cottages. As Thomas Hardy put it:

Figure 9.4   Matrix-supported clasts in the basal Lower Jurassic near Ogmore-by-Sea, Mid Glamorgan, South Wales. Photo DVA.

Only a man harrowing clods
   In a slow silent walk . . .
Only thin smoke without flame
   From the heaps of couch grass;
Yet this will go onward the same
   Though Dynasties pass.

The sudden storms, in the same metaphor, are the fall of dynasties, the drowning of Atlantis, the destruction of Pompeii.

These are the 'rare events', the brief 'happenings', the effects of which may be catastrophic both on the landscape and on its life. The small long lasting showers may leave no record at all and in places I think that all we have preserved for us in the geological record is that of the rare event.

Landslips and earthquakes

Another natural catastrophe which may be regarded as a rare event is landslipping. Thus a large part of Ramnefjell (Raven Mountain), near Loen in Norway, twice fell into Nordfjord this century, in 1906 and 1936, sending a great wave through the adjacent villages causing much loss of life and carrying a steam-ship way inland. I must admit to feeling a little nervous when we camped by the fjord in 1978 and looked up at the great scarred mountain (Figure 9.5). Many millennia of gradualism would not have done this. We have recently had the news of a great land-slip on the side of Mount Cook, New Zealand's highest mountain.

In South Wales we are only too aware of landslipping, especially the Aberfan tragedy on 21 October 1966, when a great mass of a coal tip slid down on the village destroying houses and a school, taking the lives of 128 people including 116 small children. This, of course, was man-made due to foolishness and greed in piling the waste from a nearby coal-mine on the slope of a steep-sided valley and above a spring-line. But unconsolidated sediment is always unstable.

Natural landslips are often caused by earthquakes as happened in the Alaska earthquake of 1964, when the town of Anchorage was badly damaged and whole houses in the residential quarter of Turnagain Heights were carried seawards. Earthquakes too are obviously sudden disasters that affect the Earth today and have done so repeatedly in the past. I congratulated myself for having spent three months in hospitable Japan without experiencing three well-known local phenomena: a tea ceremony (which I understand is very painful for western knees), a communal bath (which was the most difficult for a shy person like myself to avoid) and an earthquake (although there was a seismograph right outside my office in Nagoya). No doubt there were earthquakes, as there are daily in those troubled islands, but they were not severe enough for me (or that seismograph) to notice. The great Tokyo earthquake of 1923 killed about 100000 people and the city was destroyed by fire. It was rebuilt in a few months, on exactly the same pattern. The same happened after the great fire raids of the Second World War. I was told in 1981 that a quarter of Tokyo had been demolished and rebuilt within five years. An office block progressed from nothing to occupancy in the Nagoya back street where I had my apartment, while I was away for a month. But this is another of my digressions, in this case from earthquakes. We certainly saw the effects of these in several places in Japan. At Sobetsu in Hokkaido we saw a mental hospital that had been destroyed by the radiating fractures which resulted from the upward push of lava in nearby Mount Usu in 1977. At Toya Hot Spring we saw an apartment block that had been cut in half by a related fault (Figure 9.6). Several other blocks were half buried in the associated mud flows. After this eruption a seismic station was established at Hokkaido University in Sapporo and manned day and night.

Ambraseys (1971 etc.) showed how changing building styles in the past could be used as a guide to the periodicity of earthquakes, especially in the Middle East. For a while the people of classical times built their houses and temples to resist the effects of earthquakes, but when no shakes had happened in a very long time, they became blasé and did not bother. At Delphi in Greece, where the ancients went to consult the oracle about the future, a geologist can see that the future is likely to be

Figure 9.5   Scar from huge landslip from Ramnefjell (Raven Mountain) near Loen, Norway. Photo DVA.

Figure 9.6   My wife surveying a catastrophe. A block of apartments at Toya Hot Springs, Hokkaido, Japan, which had been cut in half by an earthquake associated with a local volcanic eruption four years before. Photo DVA.

catastrophic. The Temple of Athena is now only three pillars and a pediment as a result of landslipping in historic times (Figure 9.7) and nearby there is modern scree marked by the slickensides of very recent movements.

The great fault systems that run up through East Africa continue into the Middle East as far as Lebanon. It was certainly an earthquake and not the Israelites' trumpets that caused the fall of the walls of Jericho (see Chapter 12). Earthquakes today are largely restricted to plate margins such as the north of the Indian subcontinent. Some of us climbed to the earthquake damaged temple above Srinagar when we were trapped by the snow in the Vale of Kashmir during the International Geological Congress in 1964. The western edge of the North American plate will always be associated with the disastrous San Francisco earthquake of 1905 and the subsequent fire, due to a conscientious housewife cooking ham and eggs for her husband on a stove with a damaged chimney. Further catastrophes are always expected here due to movements along the San Andreas Fault and that fascinating metropolis has been called 'the city that waits to die'.

Earthquakes and landslides are very familiar and disastrous phenomena down the west coast of Central and South America. In recent years we

Figure 9.7   Partly restored temple of Athena, Delphi, Greece, which was largely destroyed in the fifth century BC by rocks crashing down from the mountain above during an earthquake and a violent storm. Photo DVA.

seem to have heard of many of them in unfortunate countries such as Mexico and Nicaragua.

We have been fortunate in Europe so far as major earthquakes are concerned, though the great Lisbon earthquake of 1755 is estimated to have killed 60000 people in that spectacular city and to have deeply disturbed the philosophy of the 'Age of Reason', including that of Voltaire's Candide. In this century, the Messina earthquake in Sicily on 15 January 1909 was said to have killed about 100000 people, but with the usual subsequent fire the casualties were subsequently estimated as twice that number.

Anyone who doubts the violence and severity of present-day natural phenomena should go to the splendid films in the 'Futuroscope' near Poitiers in west central France. There he or she will see a spectacular storm, the viciousness of a hurricane, the abrupt collapse or 'calving' at the end of a glacier, the effects of an earthquake and the sudden explosions of volcanoes, both above and below the sea. They may have them also in the various Disney creations in the States, but I have avoided these so far.

10

Evolution by accident

The day before the publication of *The Origin of Species*, Thomas Huxley wrote to Charles Darwin: 'You have loaded yourself with an unnecessary difficulty in adopting *Natura non facit saltum* so unreservedly.' (quoted in Gould & Eldredge 1977). In other words, though Huxley was the great defender of Darwin, he was criticizing him for presuming that 'Nature does not make jumps'.

One of the most fundamental changes that have happened in recent years in our thoughts about the geological past was again one towards concepts of episodicity, this time in the evolution of life. Thus the doctrine of what is called 'punctuated equilibria' replaced that of 'phyletic gradualism', which had been the almost subconscious presumption of palaeontologists since the days of Darwin. This intellectual revolution was brought about chiefly by Stephen J. Gould of Harvard and his colleague Niles Eldredge. The first epoch making paper by Eldredge & Gould (1972) put forward the idea that Nature did indeed make jumps. They maintained that evolution proceeded by short, sharp changes, punctuating long periods of *stasis*, rather than by slow progressive changes, which had been assumed since Darwin (1859) wrote of 'descent with slow and slight modifications' and the 'accumulation of successive slight favourable variations'. On the other hand Engels, that remarkable capitalist who supported Karl Marx, said that 'nature is composed entirely of leaps'. I regret to say that I prefer the view of one of the founding fathers of communism to that of one of the founding fathers of evolution by natural selection.

However, my predecessor at Swansea, Frank Rhodes, has argued (1987) that Darwin was not as gradualistic as is generally supposed. Rhodes showed that in the six successive editions of *The Origin of Species*, Darwin's views changed and he recognized the importance of the isolation and development of local varieties with the rise of new species and their migration to other areas. I shall discuss this below under allopatric speciation.

Since the ideas of Darwin and Wallace first burst upon the scientific world, we need no longer concern ourselves with the opposition of the 'fundamentalists' and 'creationists', unless we live in California (see the

'Disclaimer' in the Preface of this book). The arguments about the literal truth of the Bible is one of Wordsworth's 'battles long ago' so far as I am concerned. It is quite obvious that if we do not accept evolution, then the fossil record shows us quite clearly that we would have to accept many creations, not just the one of Genesis. This to me is just nonsense, so I am not going to waste any more time on it.

The interesting discussion now is not between 'evolutionism' and 'creationism', but between 'phyletic gradualism' and 'punctuated equilibrialism'. I think it is probable that both occurred in different groups of organisms, but that the latter was by far the more common process. Perhaps I am somewhat prejudiced by my own research in my own specialist group, but five years after their first paper, Gould & Eldredge reconsidered the matter (1977) and could only find one example which seemed to show phyletic gradualism, out of the many evolutionary studies that had been published. This was a statistical study of the Permian foraminifer *Lepidolina* from Japan (Ozawa 1975) in which, allowing for confidence limits for the mean, there was a clear gradualistic increase in the diameter of the proloculus. On the other hand a study of borehole samples of the radiolarian *Pseudocubus* through the Pleistocene (Kellogg 1975) showed a clearly marked series of steps. It has also been suggested for the Radiolaria that their evolution and extinction through the Pliocene and Pleistocene was episodic and could be related to magnetic reversals, though this theory was probably based on too few data.

Undoubtedly, as so often happens in science, both schools are right in some cases. Gould & Eldredge (1977) conceded that phyletic gradualism did occur but, like Captain Corcoran's sea-sickness in Gilbert and Sullivan's *HMS Pinafore* – 'Hardly ever'. This was also my conclusion about British hurricanes in the last chapter.

Personally, I subscribe to the views of J. G. Johnson (1982) that phyletic gradualism may have been the rule in pelagic forms and punctuated equilibrialism in members of the benthos. His own examples were from the Devonian of Nevada and he compared the presumably pelagic conodonts with the certainly benthic brachiopods. By their very nature, pelagic organisms are 'most likely to have inhabited extensive, gradually changing environments and are therefore the most likely to have evolved by a rate and a pattern that can be described as phyletic gradualism'. On the other hand 'stationary benthic organisms are the most likely to have inhabited environments that are subject to relatively abrupt changes and are therefore the most likely to have evolved by a rate and pattern that could be described as punctuated equilibria'. Certainly it has been my own experience, in my research on Mesozoic Brachiopoda, that their evolutionary patterns were usually, if not always, punctuated. I have demonstrated this in the rhynchonellid *Gibbirhynchia* and in the terebratulid *Nucleata* and its descendants (Ager 1986a). In the former case we see one species, *Gibbirhynchia muir-woodae* persisting through a considerable thickness of Lower Jurassic (Pliensbachian) strata on the Dorset coast of south-west England and then at least four different species appearing suddenly at the very top of the stage with a change in sedimentary facies (Figure 10.1).

In the latter example we see the genus *Nucleata* persisting, with very little change, from somewhere down in the Triassic right through the Jurassic and up into the Cretaceous. Only at the very end of the Jurassic and at the very beginning of the Cretaceous did it suddenly give rise to three distinctive genera, *Triangope*, *Pygope* and *Pygites* (Figure 10.2). Both *Gibbirhynchia* and *Nucleata* are good examples of *stasis*, followed by episodes of rapid evolution.

In my lectures I like to illustrate this process with two completely non-geological slides. One is of a bus stop in Greece, with the sign ϛταϛιϛ (i.e stasis); in the foreground is my mobile geological home (a motor-caravan), and in the background is Parnassus where the ancient gods did naughty things. My other *stasis* slide is a cartoon from a French magazine with the familiar white-bearded figure grabbing at a chamber-maid who says: 'En tout cas, monsieur Darwin, vous n'évoluez pas beau-coup', which I would translate as 'Anyway Mr Darwin, you haven't evolved very much'.

Gould constantly emphasized that 'stasis is data', in other words noth-ing happening in a fossil lineage is just as important as something hap-pening. One thinks of the House of Lords in another Gilbert and Sullivan work, *Iolanthe*, who 'did nothing in particular, but did it very well'.

Returning to my beloved Mesozoic brachiopods, my general im-pression in all the lineages and successions I have studied, is one of sudden changes and punctuated equilibria. Thus in Lower Jurassic (Lower Sinemurian) of Hock Cliff on the River Severn, in the west of England, there is a sudden change within a uniform alternation of limestones and shales from *Piarorhynchia juvenis* below to the related form *Cuneirhynchia oxynoti* above. Both are small rhynchonellids and both are abundant, but there is no obvious indication of any change in the environment. There is also no suggestion of a gradation between the two. So I am an unrepentant punctuationist. Nevertheless, my French friends and fellow specialists, Yves Almeras & Annick Boullier (1990) have claimed phyletic gradualism in the French Jurassic terebratulid *Caryona*. So perhaps I should not be too dogmatic.

I have never really trusted philosophers since one of them told me, with superb lack of logic, where I should take my students on geological field-trips. I therefore look somewhat askance at the views of that great philosopher Karl Popper (1979) as interpreted by Perutz (1986), when it comes to evolution. He tells us, in effect, that we should take a more positive view of evolutionary changes and should consider the prob-ability of animals moving into more favourable environments. Of course this is only possible if the organisms concerned can move and palaeontol-

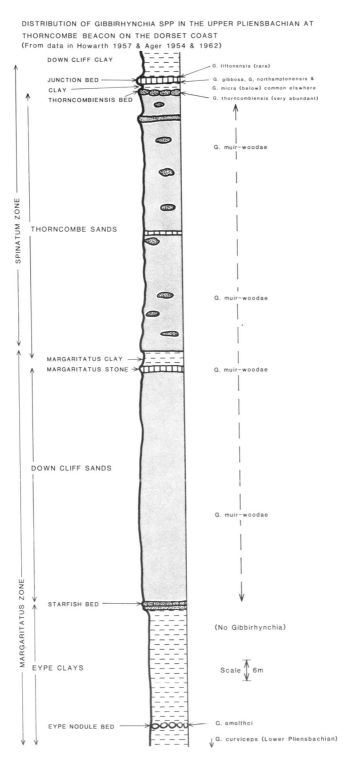

DISTRIBUTION OF GIBBIRHYNCHIA SPP IN THE UPPER PLIENSBACHIAN AT
THORNCOMBE BEACON ON THE DORSET COAST
(From data in Howarth 1957 & Ager 1954 & 1962)

DOWN CLIFF CLAY

G. tiltonensis (rare)

JUNCTION BED

G. gibbosa, G, northamptonensis &

CLAY

G. micra (below) common elswhere

THORNCOMBIENSIS BED

G. thorncombiensis (very abundant)

G. muir-woodae

SPINATUM ZONE

THORNCOMBE SANDS

G. muir-woodae

MARGARITATUS CLAY
MARGARITATUS STONE

G. muir-woodae

DOWN CLIFF SANDS

G. muir-woodae

MARGARITATUS ZONE

STARFISH BED

(No Gibbirhynchia)

EYPE CLAYS

Scale 6m

EYPE NODULE BED

G. amalthci

G. curviceps (Lower Pliensbachian)

Figure 10.1   Evolution of the brachioped *Gibbirhynchia* in the Lower Jurassic (Upper Pliensbachian) at Thorncombe Beacon on the Dorset coast in south-west England. A long period of stasis is represented by the species *G. muir-woodae*, followed by a sudden burst of four or five species in the beds at the top of the section. Drawn by Mr J. U. Edwards (Ager 1986a).

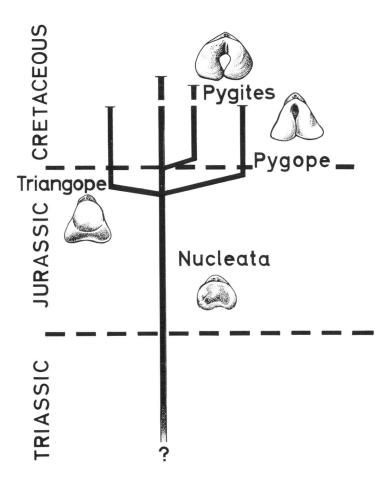

EVOLUTION OF THE PYGOPIDAE

Figure 10.2   Evolution of the pygopid brachiopods in the Mesozoic. The genus *Nucleata* remained in a state of stasis from Triassic to Cretaceous times, but suddenly evolved into three other genera at the end of the Jurassic. Drawn by Mr J. U. Edwards (Ager 1986a).

ogists have always known that mobile organisms such as conodonts, ammonoids and vertebrates evolved more rapidly than their more static contemporaries such as brachiopods and corals. That is what makes the former groups so much more useful stratigraphically.

Another obviously useful group, especially for oil geologists, is the phylum Foraminifera. My good friend Fred Banner, an eminent expert on the planktonic foraminifers, has repeatedly demonstrated to me the apparent gradualistic evolutionary pattern in many of these forms, on the basis of the many thousands of specimens available from bore-holes. This seems to be true, for example, in the evolution of the foram *Pullenia-tina*. The oil men have so many specimens and can claim a continuous record in their bore-holes. Their work may be criticized for the lack of quantification, but life is short, the economics of oil exploration are always pressing and they just have too much data. Perhaps the most convincing demonstration of gradualistic evolution was that by Malmgren & Kennett (1981) dealing with 30 000 measurements of the genus *Globorotalia* in a bore-hole core ranging through the whole Pliocene. One must always beware of confusing evolution with migration and this is particularly true in *Globorotalia*, which is known at the present day to vary markedly across the North Atlantic (Ericson, Wollin & Wollin 1954, redrawn in Ager 1963, p. 178). In that particular example, any short length of an illustrated core off north-west Africa would seem to give a clear impression of an evolutionary change in coiling direction. Since this has changed repeatedly since Pleistocene times, it would seem that it was caused by migration with repeated and sudden changes in some environmental factor. Even among the planktonic foraminifers, however, there are sudden appearances, such as the spiny globigerinacean *Hantkenina* in the Eocene.

My first predecessor at Swansea, Professor (later Sir Arthur) Trueman was responsible for one of the classics of evolutionary palaeontology, (1922). This was to demonstrate the evolution of the flat oyster, now called *Liostrea*, into the coiled oyster *Gryphaea*, in the local Lower Jurassic cliffs of Glamorgan. This became what was called the *Drosophila* of palaeontology, generating an immense number of papers. This gradualistic evolution was seemingly proved statistically, then later it was disproved statistically! Hallam (1959) rejected Trueman's hypothesis, making it all a matter of growth or allometry which Gould defined as 'the study of size and its consequences' (1966 & 1972). When I came here, my own explanation (Ager 1976, p. 136) was that the flat oysters were adapted, like their modern counterparts, to cementing themselves on hard substrates. In the Lower Jurassic of Glamorgan these were provided by chert clasts derived from the local Mississippian (Dinantian) limestones. As one goes up the succession, the chert fragments disappear and *Liostrea* is replaced by *Gryphaea* which was clearly adapted to life on a soft muddy sea-floor. Indeed, we know now that *Gryphaea* had been around for a very long time and evolution may, in fact, have gone the other way, with the cemented oysters as the 'new idea' in Early Jurassic times. So it was probably a sudden environmental change – an ecological accident – which brought about the allegedly gradualistic evolutionary lineage.

Nevertheless, it must be pointed out that nothing is simple in palaeontology. My friend and former colleague Andy Johnson has studied (with C. D. Lennon) the evolution of *Gryphaea* in the Middle Jurassic of western Europe (1990). Their conclusion in this case was that evolution was essentially gradualistic, with little evidence of stasis before or afterwards. Again this was supported by biometrics. It would be a dull world if there were no problems!

It is a curious coincidence that it was another great contribution by Arthur Trueman that made me think and talk of punctuated equilibria long before I knew that this was what it would be called. I refer to the first part of Trueman's major monograph with Weir (1946) of the non-marine bivalves in the British Pennsylvanian (Westphalian) Coal Measures. Their text-figure V (Figure 10.3 here) shows the relative abundance of the various genera, with *Carbonicola* terminating abruptly after a period of great abundance and the related genus *Anthracosia* bursting into dominance just as abruptly at the same level. Similarly *Naiadites* is replaced equally suddenly by its close relative *Anthraconauta* (Ager 1976, p. 134). So my great predecessor after first apparently proclaiming 'phyletic gradualism' then demonstrated what would now be called 'punctuated equilibria'.

So that brings me back to the 'catastrophism' which is the theme of this book. Changes have been repeated and sudden and these constitute the 'evolution by accident' which I am discussing in this chapter. I think

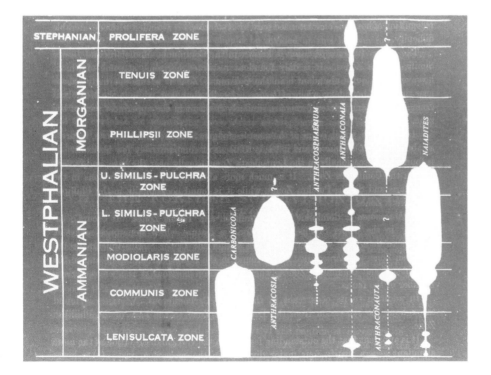

Figure 10.3   Ranges and relative abundance of genera of non-marine bivalves in the British Upper Carboniferous Coal Measures. From Trueman & Weir (1946) by kind permission of the Palaeontographical Society.

that evolution can repeatedly be shown to have resulted from accidents of three kinds: geographical, ecological and opportunistic.

Geographical accidents are those in which part or parts of a widespread species become isolated in some way and then change significantly to suit the environment available. The most famous example, of course, is provided by Darwin's finches on the Galapagos Islands off Ecuador. In this case it was both geographical isolation and the differentiation of feeding habits (from insect eating to nut cracking) that produced a number of species from a single mainland stock. There is even the finch that tries to be a woodpecker but lacks the necessary long tongue and massive beak, so pokes insects out of cracks in trees with a stick. A further sharp-beaked finch has recently been discovered on a barren island north of the main group, which drinks blood from that strange seabird the booby. Perhaps this originated from feeding on parasites. So geographical and ecological accidents can produce new forms. Another fact that has emerged quite recently is that these volcanic islands are much younger than had formerly been supposed, perhaps having formed as recently as 69 000 years ago. So evolution here must have happened very rapidly as a result of a geological accident and ecological opportunities.

I have always been even more impressed by the giant tortoises on the Galapagos, which differ from island to island and which were clearly not able to cross the intervening waters, though in this case I know of no apparent differentiation of function. However, it is a general feature of geographical isolation on islands that giant forms develop. Thus the largest tortoise of all lives on the island of Aldabra in the Indian Ocean, north of Madagascar. There is another giant tortoise on the island of Tahiti (to be seen near the home of Gauguin) and Malta had strange creatures such as a giant swan. The largest known lizard is the Komodo 'dragon' which grows up to 10 feet long (3.05 m) and is known only on Komodo and three other tiny islands in Indonesia. Instead, other islands around the world developed undersized forms such as the dwarf elephant of Sicily, the skulls of which, with their large nasal openings, may have given rise to the Cyclops legend of a one eyed giant.

A remarkable example of evolution by accident is that of the Hawaiian goose or Ne Ne (Figure 10.4). This is the rarest goose in the world and appears to have been derived from the Canada Goose, probably by some being blown off course by storms and then isolated on the Hawaiian islands to evolve in its own way. Islands seem to have been hot-houses of evolution. Presumably gigantism occurred in the absence of competition and dwarf forms may have resulted from intense competition for food. Both phenomena must result from the accident of separation from the main interbreeding populations.

Another feature of island populations is their tendency towards great diversity. This presumably results from the need for some kind of differentiation within a limited environment. A good example of this is pro-

Figure 10.4   Ne Ne or Hawaiian goose, the rarest goose in the world and a good example of 'Evolution by accident' since they probably came from Canada geese, blown off their normal routes by storms and evolved independently on the Hawaiian islands. Examples preserved in the Wildfowl & Wetlands Centre near Llanelli, South Wales. Photo Mrs Renée Ager.

vided by the island of Madagascar (which we must now remember to call Malagasy). This had nearly 40 species of lemurs, though their gentle ways made them easy to catch and therefore easy to be extinguished. There are now only 12 species left. There are also seven species of baobab (the 'upside-down' tree, which looks as though it has its roots in the air). This compares with only one species of baobab in the rest of Africa. There are no less than 450 species of frogs (compared with eight in Britain).

Lakes are sometimes comparable to islands in the diversity of their faunas. Thus Lake Malawi in the Rift Valley of Kenya has a fantastic variety of fish which, like the Galapagos finches, must have originated from a very few ancestors. It is a classic in the tropical fish world. This would be opportunistic evolution.

One of the most startling examples of evolution by accident was described recently by Furness (1988) from the Scottish islands of Foula, in the Shetlands, and Rhum, in the Hebrides. He found that sheep on Foula and deer on Rhum had taken to the habit of biting the heads, wings and legs off live seabird chicks. This unlikely habit evidently originated due to the isolation of the mammals on islands where there was a lack of calcium and phosphorus in the local vegetation, especially in the particular grasses which they choose to eat. At first the peaceful herbivores simply took to chewing old bones and antlers, but under extreme conditions, where the 'accident' involved both a deficiency of minerals in the plants and an abundance of ground-nesting seabirds, the herbivores became carnivores. No-one who has seen those wonderful wild-life films on television can doubt that Nature, in Tennyson's words, is 'red in tooth and claw'.

So our sentimentalized image of sweet Bambi-like creatures and placid sheep must be replaced by one of rapacious carnivores, not safely grazing, but red in tooth if not in claw. It is somehow obscene to imagine such a rôle change but this is clearly what may happen with the accident of geographical isolation.

It is the exact opposite of what happened to the giant panda, described by Stephen Gould in the first essay in his brilliant book *The Panda's Thumb* (1980) as essentially a herbivore 'designed' as a carnivore. The 'thumb' is a modified wrist bone which enables the animal to strip the leaves off its favourite bamboo food. Presumably its flesh-eating ancestor was isolated in bamboo forests high up in south-west China, where there was little in the way of prey. Today it is restricted to a small area at about 10 000 feet (3 600 m). It cannot stand high or low temperatures and I wonder if its black and white coloration is not a most effective camouflage in snowy coniferous forests (Ager 1991b).

It is altogether a very inefficient animal and long before Gould's book, I had been lecturing about the panda as one of Nature's accidents. I used it to illustrate the point that evolution can 'go wrong' as it were. It is thoroughly deficient in most facilities. It has poor eyesight, poor hearing and a poor sense of smell. It is slow-moving, though it can at least climb well, presumably to escape its few enemies (possibly leopards, dogs and zoo-keepers). Perhaps that is how it has managed to survive. It has a very short breeding season and the females can only become pregnant about three days a year. No doubt this explains the lack of success in mating them in zoos. It is very small and vulnerable when born; it is a very slow developer and cannot walk for months after birth.

However, its biggest problem is that of diet. It has the front teeth of a carnivore and it will eat small mammals, but it is too slow-moving for hunting, though I learned from the BBC that it can be lured with pork chops! Perhaps the 'accident' which befell that wholly flesh-eating ancestor was to find itself stranded in these high mountain areas with little to eat, whilst more efficient carnivores had taken over lower down. Its back teeth are flattened to grind a vegetable diet, but its digestive system is also inefficient in that the larger part of what it eats passes straight through without being digested. It has retained the short gut of a carnivore whilst having evolved a skeleton adapted to a herbivorous diet.

Altogether, though it is the symbol of the World Wide Fund for Nature and very 'cuddly' when seen through anthropomorphic eyes, it is really a dead loss as an animal and it is not surprising that this evolutionary accident is close to extinction. For all its attractiveness and our desire to save it to be loved by future generations, perhaps as long-sighted geologists we should say 'let it go'. Evolutionary accidents have furnished and finished many, many species in the past and certainly more will go in the future; 'Death makes room for life'. If the dinosaurs had not become extinct, there would have been no explosive evolution of the mammals and we would not be here today.

When I first wrote this, sitting in my garden, I suddenly noticed a member of the plant kingdom which seemed almost equally inefficient. These were my paeonies, which do not have the strength to hold up their heavy floral heads, but this – like some of Darwin's pigeons – is a matter of human interference and may be outside the rules of the present game.

Of course, Darwin's great lesson was that of natural selection, operating on a diverse population to favour those forms which had a particular advantage in size, strength or any other desirable quality. Besides natural selection, Darwin postulated sexual selection as a force in evolution. In this the female normally selects the male for some quality that may appear to us to be trivial, such as colour or a function-less decoration. We can hardly expect to be able to demonstrate this in the fossil record, though Gould (1974) and others did use it to explain the exaggerated antlers of the so-called 'Giant Irish Elk' (which wasn't an elk). That hardly explains its recent extinction, unless we accept such delightful notions as the thought that it could not get between the trees when the forests spread north in a warm spell at the end of the Pleistocene glaciation.

Nevertheless, sexual dimorphism has often been postulated in various fossil groups and is particularly well documented in the ammonites. In my own specialism, the Mesozoic brachiopods, it has been suggested for forms such as the Late Jurassic rhynchonellid *Torquirhynchia inconstans*, which is either right-skewed or left-skewed, like the living crossbill, *Loxia curvirostra*, which has a crossed beak. I do not know about these finches, but the asymmetry of the brachiopod is certainly about 50% each way, though one can hardly expect intermediate forms. There is no evidence, however, of this being sexual in origin (Ager 1969). I think it was simply an adaptation to enable the brachiopod to function either way up in shallow, high-energy seas.

Isolated marginal populations may well be the most common way in which new species originate. The current jargon for such stranded populations is 'peripheral isolates'. Probably my jargon is no better than the next man's, but I prefer to call the process 'allopatric speciation'. Clearly such populations, at the extreme geographical edge of the normal distribution of the species concerned are likely to be at the extreme edge of possible environments for that species and perhaps differ slightly from the main stock in the centre. They are also in greater danger of being completely isolated by some environmental change. A new ridge may be pushed up or a new valley eroded or a new seaway may form. New predators may invade the area or some vital food-stuff may disappear.

Good examples of such isolated populations at the present day are the Lake Baikal seal, *Pusa sibirica*, all by itself in the centre of Asia and the similarly isolated seal of the Caspian, *Pusa caspica*; both of these have been separated from their marine cousins in the Arctic since pre-Pleistocene times. A more recent example is the Asian lion, *Panthera leo persica*, which is only preserved in the Gir Nature Reserve in north-west India, This differs from the African lion, *Panthera leo leo*, just in details of its skull and the tufts of hair on its tail and 'elbows'. As its name implies, it was formerly found in Persia (Iran) and the last one was killed there in the 1940s. Previously there was a continuous distribution of lions round from East Africa to India, as is attested to, for instance, by the lion hunting pharoahs in Ancient Egyptian bas-reliefs. Previously they were presumably one interbreeding species, but once the two sub-species were separated, they might have evolved into two distinct non-interbreeding species.

In the fossil record I have claimed examples of allopatric speciation in my Mesozoic brachiopods. In the Lower Jurassic (Upper Pliensbachian), by far the commonest form of the zeilleriid genus *Aulacothyris* is the type species *A. resupinata*, which is found in south-west England. There was, however, a barrier across central England of ironstone depositing conditions with which zeilleriids apparently could not cope. This separated off an area in Yorkshire, in north-east England, where other members of the genus flourished. In that region we find at first (though rarely) the species *A. pyriformis* and then the more abundant related form *A. fusiformis* (Ager 1963, Figure 15.5). Yorkshire prides itself in being different from the rest of England and this evidently was so back in Early Jurassic times, when this marginal population of brachiopods evolved independently of the main stock. This was clearly an example of accidental isolation of a population leading to the evolution of new species.

Another example is provided by the Lower Jurassic rhynchonellid *Homoeorhynchia* (Ager 1983). This long-ranging stock was characterized in the Sinemurian and Pliensbachian stages by the very distinctive type species *H. acuta* but this repeatedly gave rise to slightly different marginal populations, for example in Turkey and again (twice) in Yorkshire. Most important, however, was a marginal population in Czechoslovakia called *H. maninensis*, which lived in a slightly deeper environment than the rest. With the changes that occurred at the end of Pliensbachian times, with generally deeper water over western Europe and north-west Africa, then this form gave rise to *H. meridionalis* which was to dominate the Toarcian stage that followed and replaced *H. acuta*. (Figure 10.5). This then was a matter of 'accidental' environmental change providing the opportunity for the evolution of a new species.

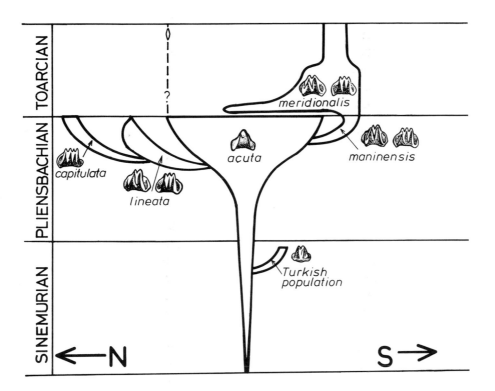

Figure 10.5   Allopatric speciation in the Early Jurassic brachiopod *Homoeorhynchia*,
showing marginal populations of the species *H. acuta* in the Sinemurian and
Pliensbachian, one of which gave rise to a new species *H. meridionalis* in the Toarcian.
From Ager (1983) by kind permission of the Palaeontological Association.

Earlier, J. G. Johnson (1975) had postulated allopatric speciation in
no less than 16 species groups of Devonian brachiopods, whilst phyletic
evolution was 'not evident'. These were all North American forms ranging from California and Nevada to the Arctic islands.

Probably the best modern demonstration of speciation in the Recent
fossil record has been the work of Peter Williamson (at Bristol and
then also in that hot-house of evolutionary thought in the Museum of
Comparative Zoology at Harvard). In a series of papers (1981a, b,
1982a, b etc.) he documented the evolutionary changes in a number of
different stocks of Cenozoic molluscs in some 400 m of sediments on
the east side of Lake Turkana in northern Kenya. The material is very
abundant and the record shows clearly that abrupt changes occurred
simultaneously in many different lineages, producing new taxa. These
happened particularly when there were sudden changes in lake level and
at particular levels of tuff fall (Figure 10.6). The lack of intermediate
forms between the taxa is noteworthy and there is no evidence whatever
of gradualistic changes.

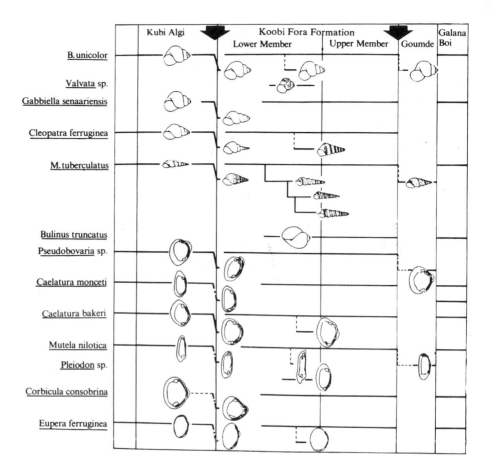

Figure 10.6 Punctuated evolution in Caenozoic freshwater molluscs in the Turkana Basin, Kenya. The abrupt appearance of new forms (simultaneously in many groups) coincides with changes in low levels of the lake as indicated by arrows. From Williamson 1981a, by kind permission of the author and © Macmillan Magazines Ltd.

Abrupt changes in lake level are widespread in Africa south of the Sahara and reflect climatic changes in Late Pleistocene and Holocene times. 'Pluvials', with heavy rainfall, may be correlated with the 'glacials' of higher latitudes. Lake level changes often occurred so fast that they cannot be resolved by radio-carbon dating (Street-Perrott & Perrott, personal communication).

All evolution must depend on the diversity of the basic stock from which new species and subspecies arise. One has only to study the variation in the pigeons in Trafalgar Square, London or those in the Piazza of San Marco in Venice to appreciate the wide variety of forms (e.g. from pure black to pure white) in what are obviously single interbreeding populations. If one considers the variation within a species throughout its range of adaptations and geographical distribution then the contrasts may be even more marked. The obvious example here is the dog, *Canis familiaris*, which may have been domesticated 8000 or more years ago and has been highly selected by man for various functions. This may not be 'natural' evolution, but it shows what is possible. The dog probably came originally from the northern wolf (possibly with some admixture of jackal), but no species shows a greater variety of 'breeds', perhaps 400 world-wide. It is difficult to imagine a pekinese breeding with a mastiff, but presumably it is possible since they are the same species. The range of variation in the dog is almost unbelievable. Figure 10.7 shows the chihuahua 'Blo' belonging to our friends Pam and David Cullen in New Zealand, wearing the collar of his predecessor, 'Ptolemy', a Great Dane, to show the possible size variation within a species.

Figure 10.7 'Blo' – a chihuahua which belonged to our friends Dr David and Mrs Pam Cullen in New Zealand, wearing the collar of his predecessor 'Ptolemy' a Great Dane, to show the possible range of variation within a species (in this case *Canis familiaris*). Photo Mrs P. Cullen.

The range of variation within the dog has, of course, resulted from selective breeding by people, who wanted dogs for different purposes. Thus there are dogs such as the greyhound selected for their great speed, there are dogs such as the pointer which hunt by sight, those such as the bloodhound which hunt by smell, there are the retrievers which bring back shot birds, there are the terriers which dig and the water spaniels which swim. The bull-dogs and bull-terriers were probably selected for their powers of holding on in the brutal 'sport' of bull-baiting (thereby foreshortening their jaws and giving them respiratory problems). I must admit to having been beaten when it came to finding a function for the so-called 'lap dogs', until I read the suggestion that ladies used to hold them on their laps in the hope that the warmth of the dogs would attract the fleas from their own bodies! So every breed of dog has its particular function.

In our clinical world we do not always appreciate the importance of parasites on mankind, though one only had to talk to a soldier of the First World War to realize what it must have been like in the past. I am always struck by the record that when Thomas à Becket fell murdered to the ground in Canterbury Cathedral, swarms of lice fled from his cooling body. There are internal parasites too, as one should realize when one sees pot-bellied African children on television. Every species has its parasites and this apparently always was so (Ager 1963, Chapter 15). It is a fundamental part of the process of evolution and the extinction of the host species must always have resulted in the extinction of the parasites which depended on it. For them it was a most unfortunate accident.

Returning to the question of speciation, as a palaeontologist it has always puzzled me why *Homo sapiens*, unlike other successful taxa in the history of life on Earth, has not split and diversified, though admittedly the record of that particular species, in geological terms, has been very short. To evolve new forms, the basic stock has to be split up and different populations have to be isolated in some way, becoming separate subspecies and, in due course, separate species which do not interbreed.

Perhaps we can see the beginnings of this in man in the abominations of nationalism and racialism. The isolation may be geographical (as in the mountains of New Guinea), it may be skin colour (as in South Africa), it may be religion (as in Northern Ireland) or it may be language (as in Belgium).

A South Sea islander is never likely to meet an eskimo, let alone breed with one, though it is certainly not impossible. There are plenty of so-called 'half-breeds' in countries such as Brazil. Without the huge human migrations in the past and the ease of modern travel, the isolated human populations would have gone their separate ways, becoming more and more adapted to their different habitats and modes of life. The South Sea islander would have become better adapted to a semi-aquatic mode of life, the eskimos would have become better adapted to his life in an extremely cold environment. No-one could ever accuse me of racial prejudice, but I would never back a team of eskimos in an international basketball tournament! Differences exist, (in this case, height) and there is no point in denying them. That such differences are often related to environment and habit is shown by the Japanese, who have increased in average height from about 156 to 170 cm in the present century, though their average eyesight has deteriorated.

Once there is a difference, then interbreeding is less likely to occur and the stocks slowly separate. Female selection of a mate is probably important in this connection. Tall girls rarely choose short boys. Much of the unforgivable bitterness of racialism is probably due to the nonsensical horror felt in some societies about, for example, a white woman consorting with a black man. This was the cause of much of the hatred (and many of the lynchings) in the southern United States. Without justifying it in any way, I see it as part of a trend towards subspeciation. Fortunately for our species that trend has now been reversed and I can only see benefit in the diversity in *Homo sapiens*. The subtle differences within the species are illustrated by the fact that, although there are few variations possible in the human visage, one can immediately recognize a friend's face at the other end of the Earth out of its five thousand million inhabitants.

It has been estimated that there are some 2000 to 4000 different languages spoken in the world today, though it is often difficult to distinguish between a language and a dialect. A different language means a break-down of communication and a tendency to become more and more separate. Thus Belgium's biggest problem is bringing together the French-speaking Walloons in the south and the Dutch-speaking Flems in the north. I have made myself both popular and unpopular in my present homeland by foretelling the inevitable extinction of the Welsh language. It is sad, but much greater languages have become extinct in the past. (That will probably get me into trouble again!)

Religion, for all its good intentions, has always been a divisive force in mankind. Obvious examples are seen in Northern Ireland and Israel/Palestine. Even our own Civil War in Britain was partly a matter of religious differences.

Once divided in some such way, then trouble always seems to follow, as it did with dreadful slaughter when India and Pakistan became separate countries. This happens especially in connection with the occupation of territory, as between the Israelis and the Palestinians. We see it often in the natural world in those marvellous television films. Every nation makes fun of some other nation as the British do of the Irish, the Americans of the Poles, the Romanians of the Hungarians, the Greeks of the Turks. Even within countries there are such differences, as between the Castillians and the Andalucians in Spain; northern Italy is almost a separate country from southern Italy (and not so long ago spoke what were virtually different languages). We all have joke cities and towns such as Aberdeen, with its allegedly mean Aberdonians. The Bulgarians say the same about Gavrovo, where allegedly, at festival times, the inhabitants dance in their socks, so that they can use the music from the next town for nothing!

All these digressions are merely to illustrate my point that species do become differentiated within themselves and this may lead, within the vastness of geological time, to the evolution of new subspecies and species. It may not be straining the evidence too far to suggest that the Olduvai Gorge and neighbouring areas in East Africa were an ecological 'island' where the right environment and the right anthropoid stock were available for the origin of *Homo* and all its branches. Indeed when the perverted racialist suggest that our black brothers and sisters should go back to Africa or the Caribbean, that the Chinese of south-east Asia should go back to China or that the English immigrants, such as myself, should leave Wales, then I cannot but extend that line of argument to its logical conclusion and send us all back to the Olduvai Gorge. It would be very crowded!

However, so far as our own species is concerned, as I said before, I am optimistic. I can see the various races coming more and more together. We shall all be a sort of khaki colour, all with the same religion (or none) and, unless the Chinese beat us to it, all speaking English, albeit sadly with an American accent.

The episodic nature of the biological record can hardly be better illustrated than by human history itself. The period from the first hand made silica chips in East Africa to the silicon chips of the last few decades lasted perhaps some four million years. More than 99% of human history was passed in the Stone Age, covering the period from the first human footprints found by Mary Leakey (1987) in northern Tanzania to the beginning of the Bronze Age. Thereafter technology advanced at an ever increasing rate. The siege of Troy happened, in my opinion for sordid geological reasons (Ager 1985c) near the end of the Bronze Age about 3000 years ago. The Iron Age culture had already developed to a remarkable degree of technology and artistry by the time of the salt miners at Hallstatt in Austria, about 1000 BC. Then it went on accelerating with the wheel, the stirrup and the alphabet until the iron bridge across the River Severn in the west of England marked the start of the industrial revolution with steam engines and factories and popular education.

So, like all other forms of evolution, technological history was marked by sudden spurts. My mother turned in the street to stare at the first motor-car she had seen in London. She lived long enough for me to rouse her from her bed in the middle of the night, to come down to see on television the first men landing on the moon.

However, it should be noted (as pointed out again by Gould 1980) that the social evolution of man has not been Darwinian, but rather goes back to the earlier evolutionary thought of Lamarck with the inheritance of acquired characters, in that each generation has inherited the knowledge and skills (if not always the wisdom) of previous generations.

One of the failings of palaeontological studies of evolution is that specialists tend to expect to find complete evolutionary sequences in their own backyards. This is true whether we are talking about Trueman and the oysters in the Lower Jurassic cliffs of Glamorgan or many modern micro-palaeontologists studying the forams in their favourite bore-hole. Unfortunately life (ancient and modern) is not as simple as that. We always have to be aware of migration and we have to spread our net to catch our fossils as widely as possible. In one of my earliest papers (Ager 1956, Figure 1) I showed how a variable population migrating past a given point could give the impression of evolutionary change.

Later, with two research students, Alan Childs and David Pearson, I attempted to unravel the evolutionary pattern of all the nominal genera of Mesozoic rhynchonellid brachiopods (Ager, Childs & Pearson 1972). It soon became obvious that we could not limit ourselves geographically. Some of the stocks were limited in distribution, notably to the 'Tethyan' region, others moved around the world with time. Yet others seemed insoluble and we had to leave gaps in our charts. An interesting example is provided by the gap in our dendrogram for the supposed Mesozoic members of the family Pugnacidae (*op. cit.* Figure 3); we left a gap between the Bajocian form *Stolmorhynchia* and the Oxfordian form *Lacunosella*. That gap was subsequently filled by Dr Carlo Sturani of Turin, who found *Stolmorhynchia* in the Bathonian of the Basses Alpes (personal communication 1974) and Dr. Shi Xiao-ying who visited me from Beijing in 1988 with examples of a form from China which also filled this particular gap. This shows how international one must be in such studies. One has to study evolving lineages on a world-wide basis and not be content (again like Voltaire's Candide) just to cultivate one's own garden.

In referring to our 1972 paper (see above) I might mention the clear picture that emerged of repeated 'bursts' of evolution (see Figure 10.8). First there were the Dimerellacea in the Triassic, then the Tetrarhynchiinae (twice) in the Jurassic and finally the Cyclothyridinae, originating in the Mid Jurassic and then, 'exploding', in the Cretaceous. More detail can be found in the paper, but unfortunately there are many printers' errors because the authors did not receive proofs to correct.

Ancestors are often overlooked. They are minor, rare forms which do not come to the attention of those who are studying their descendants. Some palaeontologists talk of 'faunal breaks' as if new forms suddenly appeared without mothers or fathers. All lineages must be complete even if the evolution of new forms, as I am saying here, commonly happened very quickly. The Boeing 747 did not evolve from the giant Lancasters or 'Flying Fortresses' of the 1940s, but from the first tiny jet planes, the Heinkel He178 of 1939 or from the Gloster–Whittle of 1941. These will not be found when the modern sediments of the North Sea come to be studied in the distant future. The most likely planes to be found

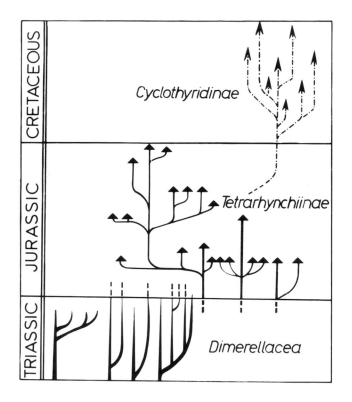

Figure 10.8   Bursts of evolution in the Mesozoic rhynchonellid brachiopods. Each separate branch represents a named genus. From data in Ager, Childs & Pearson (1972).

there are the Dakotas and the Blenheims which were vastly more abundant.

Bursts of evolution or 'explosive evolution', as it has been called, are very familiar to palaeontologists and probably deserves more attention than those overworked 'mass extinctions'. It occurs in almost every group and must have resulted from some kind of ecological accident when the group concerned suddenly had the opportunity to diversify. The most familiar example is that of the mammals at the beginning of the Caenozoic. After millions of years in a state of stasis, all through the Mesozoic, they were suddenly given the opportunity to fill the ecological niches left empty by the extinction of the large reptiles. Again 'Death made room for life.'

Personally, I am particularly impressed by the rapidity, early in Tertiary times, of the evolution of the whales. Already by Mid Eocene times they had acquired the highly specialized skulls suited to their particular newly acquired mode of life. Those skulls were elongated with teeth adapted for fish-eating and with the nostrils moving to the top of the head. Presumably the whales took over from the large marine reptiles, which became extinct at the end of the Cretaceous with what seems to have been the collapse of the whole marine ecosystem. In the bank of the Waipara River, near Christchurch, New Zealand, Dr Alexa Cameron showed me the actual point in the section where one of the last plesiosaurs had been extracted, less than a metre below the Cretaceous–Tertiary boundary (Figure 13.6). This almost made me believe in mass extinctions! But more of that later.

The most unlikely accident of all in the history of life on Earth was the origin of life itself. Someone has compared the likelihood of the right chemicals and the right forces coming together in the primaeval soup to the likelihood of a hurricane passing through a junkyard and blowing the pieces together to form a jumbo jet. But I do not feel that sort of scepticism.

Alternatively, the arrival of life on Earth as a celestial hitch-hiker riding on a meteorite, as has been suggested by several workers, simply puts the blame back on an earlier accident elsewhere. What is more, it has been calculated that organic molecules carried within comets could not have survived impacts at velocities of more than about 10 km per second. Nevertheless, it has been suggested that such molecules could have survived impacts at comparatively low velocities in the much denser Precambrian atmosphere when the main extra-terrestrial bombardments occurred.

I suppose I had better mention the concept of a divine creator, but personally I do not find that particular hypothesis useful and I am tempted to ask about the cosmic accident that created Him (presumably before the 'big bangs' that started the universe). And what did He do before He created the world and mankind?

Personally, given the resources of geological time, I feel confident that sooner or later that hypothetical chimpanzee, sitting at a typewriter, will one day type *Hamlet*. A very significant remark, made by Maynard Smith (1981) was that 'a new species arising in 50 000 years . . . is sudden to a palaeontologist but gradual to a geneticist'.

I was impressed by the studies made after the 1980 eruption of Mount St Helen's in Washington State, which destroyed all life for many kilometres around. Within a remarkably short time, nasty hot, evil-looking pools around the volcano were teeming with life in the form of bacteria and blue-green algae. These are exactly the kinds of organisms that we know from the earliest records of life on Earth. The necessary original formula must have been one of chemistry and heat in a watery environment.

Thereafter there was an excessively long period of stasis with life only represented by the simplest kinds of plants before the development of nuclei or sex. Only towards the end of Precambrian times was there the evolution of a number of different life forms, as seen in the *Ediacara* assemblage of South Australia and now known in other parts of the world (including South Wales).

However, the big evolutionary explosion came with the Cambrian and notably with the Mid Cambrian and the remarkable Burgess Shale fauna of the Canadian Rockies and elsewhere. Here we have a near miraculous preservation of soft-bodied creatures belonging to many different animal phyla. This was an accident of preservation rather than of evolution. Later miracles of this kind are provided by the Devonian Hunsrück Schiefer of the Rhineland, the Pennsylvanian of Mazon Creek, Illinois, the Jurassic of Holzmaden and Solnhofen in Germany and the Upper Cretaceous of Lebanon.

The accident of preservation is all-important and often not sufficiently appreciated, but I have already discussed the gamble and lies of preservation in Chapter 2. Not only does the record lie, but we lie about it. We tell our students that most fossils are the remains of animals with hard parts, but in fact the remains of soft-bodied animals, in the form of trace fossils, are probably much more common than the others, though it is usually impossible to count them.

Obviously the episodicity of the fossil record may be no more than a reflection of the episodicity of the sediments which contain them and no-one places more emphasis on gaps in the record than I do myself (Ager 1981a, Chapter 3). But the essential point, as I have said before, is to get out of the back-yard and look at the world-wide record. I have no time for the modern doctrine of cladistics, which is essentially a biologist's approach to evolution, looks only at shared characters of living forms and ignores the fossil record. It reminds me of the existentialist philosophy of Jean-Paul Sartre and others, in which one only really exists when making choices. So new species are only thought to originate by division and not by descent. There is a great danger of subjective judgements on what is 'primitive' and what is 'derived'. Thus a cow and a lung-fish both have backbones, which is said to be a 'primitive' character, but they also have lungs which is a 'derived' character. This leads one to the ludicrous conclusion that the lung-fish is closer to the cow than to the cod. The devotee of cladism tends to ignore the levels of development recorded in the fossil record, for example the diversification of the fish in the Devonian. Similarly I am sceptical of numerical taxonomy, which tends to give equal weight to all characters, so that the odd pimple on a trilobite may be treated as of equal importance to the form of its head-shield.

One of the most interesting aspects of the rare event hypothesis, at least to me, is what it implies in the field of biogeography. There has been much discussion of the ways in which plants and animals of land or shallow marine habitats managed to cross wide oceans. For modern plants it is not too difficult, with accidental transport by birds or floating seeds carried on ocean currents. The abundance of stranded coconuts along almost any tropical beach tells us how coconut palms became so widespread on the remote islands of the oceanic tropics. Obviously there was no conscious effort involved in that case. But when I consider our own species, I am lost in wonder at the Polynesian people who managed, in their open canoes, to reach every habitable island in the Pacific before Christopher Columbus set sail in his much larger ships to supposedly 'discover' America. The Polynesians had no maps other than bamboo splinters bound together with cowrie shells as islands. These indicated not geography in our sense, but the direction of winds and currents. One can only shudder at the thought of how many of them failed to reach the dreamed-of landfall and one can understand why some of them turned to cannibalism. It was surely a rare event when the first Polynesians reached Tahiti or the first Melanesians reached Fiji.

There has been much said about 'island hopping' to explain the distribution of certain land tetrapods. More recently the possibility of 'seamount hopping' has been proposed to explain the distribution of certain shallow water sessile benthos. Olga Zezina (1985) used it to explain the spread of modern brachiopods and I have suggested it (Ager 1986c) as the only possible alternative to an expanding Earth to show how some of my Mesozoic brachiopods managed to cross the Pacific at low latitudes when it was much wider than it is today. The dredging of Cretaceous rudistid bivalves from a sea-mount in the middle of the Pacific was clear proof that this had happened. Such 'sweepstake' migrations, as they have been called, are surely clear examples of the importance of the rare event. They may even be called catastrophic at times, as when a new arrival upsets a whole ecosystem. In Britain the so-called 'slipper limpet', *Crepidula fornicata* from America, became the commonest shell on the south coast within a few years (Ager 1963, pp. 167–8). Similarly the Japanese gastropod *Ocenebra* is taking over on the west coast of North America.

Summing up it may be said that the course of evolution was largely a matter of opportunity, brought about by an 'accidental' change in geography or in the environment or the extinction of an organism occupying a particular ecological niche, making room for a new species. I think that in most or many groups it happened in 'jerks'. Of course, the most accidental of all processes in evolution, and perhaps the most important, is that of seemingly random mutation. Such mutations are sometimes deleterious and may be due to some external agency such as ultra-violet radiation, or they may, apparently, be completely spontaneous. Whatever the cause or effect, they can only be described as accidental.

Darwin repeatedly apologized for the inadequacy of the fossil record and palaeontologists have continued to do so ever since, but I think they are unduly pessimistic. It is obviously inadequate if we are concerned with some popular group such as the birds, whose skeletons make their preservation highly unlikely, but the pessimism is not so justified if we consider the less showy marine invertebrates, especially if we consider them on a world-wide basis and not just in our own back-yards. Evolution happened, there is no doubt about it. It was usually episodic and rarely in one place. We must always be careful to distinguish between evolution and migration (Ager 1956). There still remain many problems to be solved but the solving of them is an exciting task.

II

Periodic plutonism

At the beginning of Letter 65 in his famous book on *The Natural History of Selborne*, written in 1788, Gilbert White wrote as follows:

The summer of the year 1783 was an amazing and portentous one, and full of horrible phenomena; for, besides the alarming meteors and tremendous thunder-storms that affrighted and distressed the different counties of this kingdom, the peculiar haze, or smoky fog, that prevailed for many weeks in this island, and in every part of Europe, and even beyond its limits, was a most extraordinary appearance, unlike anything known within the memory of man. By my journal I find that I had noticed this strange occurrence from June 23rd to July 20th inclusive. . . . The sun, at noon, looked as blank as a clouded moon, and shed a rust-coloured ferruginous light on the ground, and floors of rooms; but was particularly lurid and blood-coloured at rising and setting. . . . The country people began to look with a superstitious awe at the red, louring aspect of the sun; and indeed there was reason for the most enlightened person to be apprehensive. . . .

What this remarkable pioneer student of nature did not know was that all these phenomena (apart from the meteors) were certainly the effects of the eruption of the volcano Laki on Iceland, which occurred precisely at this time (P. Francis, personal communication 1990). At the same time Benjamin Franklin, writing from Paris, described 'a constant fog over all Europe and a great part of North America' (Pearce 1991). Similar effects were feared from the burning of oil wells in the Middle East but proved unfounded and Pinatubo volcano in the Phillipines had a much greater effect.

Though igneous petrology is certainly not my line, I have been fascinated since a child by such stories of great volcanic eruptions. I read Lord Lytton's *The Last Days of Pompeii* several times, with its imaginative reconstruction of the destruction of that hot Roman city by the eruption of Vesuvius on 24 August, AD 79. The excitement came through the heavy early nineteenth century style. Later in life I made a special effort three times to visit the miraculously preserved streets and houses. I also visited the remains of Herculaneum, in some ways even better preserved. It always worried me that Pliny, who wrote an account of the eruption, did not mention its most dramatic effect, the destruction and burial of these cities. When my brother fought that way with the British 8th Army in 1944, the ash was falling from the eruption that year of Vesuvius, symbolic perhaps of the twilight of Hitler and his Germanic gods.

We are always told of the wonderful sunsets around the world caused by the dust in the atmosphere from the spectacular explosion of Krakatoa, between Java and Sumatra, in August 1883, with the complete destruction of that island. It is said that the dust from this eruption caused a 20% reduction in the intensity of sunlight in France for several months. Nearly 100 years later I was excited to pick up, on the east coast of Africa, pebbles of pumice from Krakatoa which had floated all the way across the Indian Ocean. I was told that it took six months for the first specimens to make that journey.

In a more morbid frame of mind, I was also captured by the story of the eruption of Mont Pelée on Martinique on 8 May 1902, which killed nearly 40 000 people in the town of Saint Pierre, one of only two men to escape being a man in a subterranean jail, charged with murder. That always said something to me about divine justice!

Latacunga in Ecuador has been destroyed six times by lahars (hot mud avalanches caused by eruptions melting an ice-cap). These came from Cotopaxi, some 29 km away. In 1970 the town of Yungay in Peru was destroyed with the loss of 20 000 lives by such an avalanche, which travelled at the almost incredible speed of 480 km per hour (Andrews 1982). Other records speak of volcanic avalanches travelling at speeds of 400 km per hour for up to 150 km.

As a geologist I always emphasize the story of the explosive eruption of Santorini in the Aegean, about 1450 BC, which may have brought to an end the first European civilization on the island of Crete, some 60 km away. Pedants want to call the island Thira, but Santorini is a much more romantic name. They have long argued about the exact date, but I am pleased that Manning has confirmed (1989) that the eruption and the destruction did in fact coincide; it is too good a story to contradict and I like to show slides of parts of the royal palace at Knossos in Crete, buried in what seems to be volcanic debris.

I made this the final point of my account of the geology of Europe (1980), so that my history ended with a bang. Flying back once from Kenya or Tanzania, I could hardly believe my luck in spotting, far below, the unmistakeable circular outline of Santorini (Figure 11.1). I have always wanted to visit Santorini, but now my health makes it difficult and my lone-sailor daughter has beaten me to it.

All this indicates the frighteningly episodic nature of volcanic activity and no-one living in an area that has seen eruptions in geologically recent times can be sure that their seemingly slumbering neighbour will not suddenly burst into violent activity again. As Michael Andrews wrote of Arequipa (1982, p. 57); this is 'a city in Peru beneath another, young

Figure 11.1   The Greek island of Santorini (or Thira) showing the remains of the great volcano, seen from the air. Photo DVA.

active volcano El Misti, is built from it, and is famous for the beauty of its architecture . . . but on a world list of cities at risk from volcanoes it must stand high.' No doubt we are safe in Britain, where our last volcanism ended some 45 million years ago, but I sometimes wonder about Bath, where hot water continues to spurt from the ground, making this a spa town since the days of the Romans. It strikes me that the beautiful Regency town stands above a continuation of the Worcester Graben, which subsided and accumulated sediment through Early Mesozoic times. This is one of a series of graben across western Europe most of which, like the Rhine and the Rhône, have a record of geologically recent volcanic activity (Ager 1980, p. 254 *et seq.*).

As was pointed out to me by my friend Peter Hooper of Washington State University in Pullman, those who study igneous rocks do not have the same subconscious presumptions of continuity as do many stratigraphers and sedimentologists. They expect their phenomena to be episodic. One can travel for hundreds of miles over the huge staircases of the Columbia River Basalts or the Snake River Basalts of the American north-west (Figure 11.2). I spent several months as a soldier on the Deccan 'Traps' of India. In such places one cannot doubt the episodicity of the phenomena displayed. Hooper calculated (1982) that the Columbia River Basalts cover an area of 200 000 km^2 with an average thickness of more than 1 km. This filled an inland sea, the Columbia Basin, which had formed here some 40 to 60 million years ago.

Figure 11.2   Step-like weathering of the Columbia River basalts, near Vantage, Washington State, USA. Photo DVA.

The basalts, some 1500 metres thick, cover most of the eastern part of Washington State and large parts of the neighbouring states of Oregon and Idaho. All this happened between about 17 million and 14.5 million years ago, during a comparatively short episode in Mid Miocene times. Hooper also estimated that one well-known basalt, the Roza Flow, moved over 300 km in about seven days at an average speed of about 5 km per hour. Some of the flows passed through the rising Cascade Mountains to reach the Pacific.

Farther south, in Arizona, one sees vast spreads of black basalt from the air (Figure 11.3). These too date from way back in the Tertiary, but have stayed remarkably fresh thanks to the arid climate of the American south-west. The most recent eruption here was of Sunset Crater, just north of Flagstaff, whose last eruption (dated by tree rings) was AD 1064, two years before the most familiar date in British history.

To the specialist in igneous rocks, volcanic eruptions are always 'rare events'. Apart from the lava themselves, with their slowly cooled, often columnal bases and their quickly cooled tops, there is the record of the interbasaltic layers. There is repeatedly the red 'bole' of subaerial erosion on the top of lavas. This is well seen, for example, in the Carboniferous lavas of Edinburgh in Scotland and the Triassic basalt of southern Morocco (see Figure 4.3).

Figure 11.3   Huge spread of basalt below sandstone escarpment, seen from the air over northern Arizona, USA. Photo DVA.

There are also organic remains preserved between the lava flows, such as the trees in the *Ginkgo* Petrified Forest near Vantage in Washington State (from which came the petrified *Sequioa* wood of the pen-stand on my desk). In Britain we have the inter-basaltic Eocene leaf bed on Mull in the Inner Hebrides off the west coast of Scotland.

All these lava came up mostly from fissures and, being mainly basic in composition and of low viscosity, they travelled very fast. Peter Hooper commented (personal communication 1987) on the amazing speed at which the Columbia River basalts must have travelled over great distances before they solidified. Explosive eruptions such as Krakatoa, Santorini and Taupo in New Zealand, scattered vast quantities of volcanic debris over huge areas. Taupo, which erupted in AD 186 covered the whole of North Island in ash and its crater is filled with a large lake. In my wanderings around Europe and beyond, I have frequently been struck by how recently major volcanism has ceased. This is true, for example, in the Auvergne district of central France. If one stands on the top of the Puy de Dôme near Clermont Ferrand, one can see a line of volcanoes (Ager 1981a, Figure 8.2). These head off in both directions presumably along a fault-line. This is La Chaine des Puys and some of them, such as the Puy de Parion are quite perfect cones with craters in their summits. Clearly, though largely composed of soft ashes, they have not been there long enough to suffer any appreciable erosion. In the same way, volcanism only seems to have ceased yesterday, geologically speaking, in areas such as the Eifel in Germany, around Lake Balaton in Hungary (sometimes called the 'Hungarian Auvergne') where the supposed volcanic caldera of the Tihany peninsula projects out into the lake and is surrounded by cones of 'geyserite' left by geysers and recent hydrothermal activity. In Czechoslovakia there is the 'Czech Auvergne' north-west of Prague, with a landscape dotted with ash cones and volcanic necks. One can almost hear the splash where volcanic bombs landed in soft, now brightly coloured Tertiary muds and tuffs (Figure 11.4). Subterranean heat still provides the hot water of the old spa towns of Marian Lazne (formerly Marianbad) and Karlovy Vary (formerly Karlsbad). These were the haunts of the European aristocracy in years past, but more recently were reserved for deserving workers. Steam still comes from the ground and blocks the local pipes with siliceous deposits, it also preserved the perfect silicified rose given me by a Czech friend and standing on a bookcase in my study. It is the same in the large volcanic area within the Carpathian bend in Romania, in the Calimani and Harghita Mountains, and it is the same in Bulgaria. It may well be that early

Figure 11.4   Volcanic bomb in Tertiary tuff, with radiating cooling cracks, Hnojnice, Czechoslovakia. Photo DVA.

humans first kindled their fires from the conflagrations caused by such geologically recent eruptions (Ager 1980, pp. 189, 192, 249).

In Italy, the volcanoes of Vesuvius and Etna are famous, as are those of the Lipari Islands such as Stromboli (again my daughter beat me to it) and what may be called the 'type' volcano of the island of Vulcano. There were much greater happenings here in the not very distant geological past. West of Vesuvius and Naples are the Phlegraean Fields with the bubbling mud and sulphur of Solfatara, the crater of Lake Avernus (thought by the ancients to be the entrance to the underworld) and the perfect cone of Monte Nuovo, which appeared overnight on 29 September 1538 and was described in detail by Sir William Hamilton (1774) in one of the first scientific works on volcanology (when he might have been paying more attention to his wife Emma).

Farther north in Italy are the volcanic areas around Rome where the porous rocks produce the sweet–dry Frascati wines. North again is the area around Lardarello in the northern Apennines where there are white and shifting fumaroles, known and used since Roman times. The steam is now used to power the generators which supply electricity to the Italian railways. But again volcanicity was vastly more intense in Late Caenozoic times.

I had the same impression of volcanic rocks hardly cooled in Kenya where, for example, the last eruption of Mount Shetani (or Satan) is still recorded in local folk memory, though there is no active volcanism today anywhere near this locality. In other places where volcanism continues, it is nothing compared with what it once was. Thus in Japan, Mount Sakurajima still puffs away gently in Kagoshima Bay and visitors have themselves buried in warm volcanic sand at Ibusuki on the nearby Satsuma Peninsula for its supposed benefit to their health.

Mount Aso, on the same island of Kyushu, is said to be the largest active volcano in the world, 24 km across. The marginal peak of Daikambo still bubbles and puffs away in its lake-filled crater (Figure 11.5) and is much visited by parties of Japanese tourists, led by little flags. There are concrete shelters for visitors around the rim in case it gets angry again. The Japanese are the most safety conscious people I know. Nevertheless 60 000 people live on the floor of the main crater.

The speed at which volcanoes can form is illustrated by Showa Shinzan in Hokkaido, which suddenly rose as a peak in 1944 and 1945. A local amateur geologist – a postmaster – put his chin on a rest and plotted the rapidly changing profile of the volcano relative to a stretched line. They kept quiet about it during the war as they were afraid it would lead bombers to the nearby steel town of Muroran.

That most famous of Japanese volcanoes, Fuji Yama, is all at peace and the ultimate symbol of gentle Japanese art. It hardly needs illustration here since it has been painted and photographed so often. A large picture of it hangs in our hall, complete with cherry blossom. The school-

Figure 11.5   Daikambo crater on the margin of Mount Aso, said to be the largest active volcano in the world, Kyushu, Japan. Photo Mrs Renée Ager.

children at its foot sing a lovely little song about Fujisakura – the special kind of cherry blossom that only grows on that classic mountain. But the last major eruption of Mount Fuji must have been a violent and terrifying affair for the early inhabitants as seen in the reconstruction in the museum at its foot.

Similarly in New Zealand, there is active volcanicity, such as Ngauru in North Island which is continuously erupting in a not very serious way. But for the most part it is in the form of geysers, steam and bubbling mud-pools, notably in the famous volcanic area at Rotorua, where one can enjoy a thermal bath. Natural steam runs the turbines at Wairaki and comes up everywhere thereabouts as puffs of steam in back gardens. We were near Mount Tarawera in 1986 exactly 100 years after its eruption (Figure 11.6) which destroyed the silica pink and white terraces in Lake Rotorua (then New Zealand's greatest tourist attraction). It also buried three Maori villages, one of which, Wairoa, is now re-emerging as a sort of Southern Hemisphere Pompeii (Figure 11.7).

Lake Rotorua is certainly the site of a vast collapsed caldera as is what was probably the greatest of them all, Lake Taupo, also in North Island. More than 15 000 cubic metres of lava, ignimbrite and tephra are thought to have been erupted from the Taupo area, with ashes covering virtually the whole of North Island, draping the pre-existing topography to a depth of up to 10 m (Lillie 1980). All this is dated as less than 40 000 years ago. If it happened again today it would probably wipe out

Figure 11.6   Mount Tarawera near the Bay of Plenty, North Island, New Zealand, which erupted disastrously in 1886. Photo DVA.

Figure 11.7   A building in the village of Wairoa, buried by the eruption of Tarawera, North Island, New Zealand, now being re-excavated. Photo DVA.

all life on the North Island, including that in the two biggest cities, Auckland and Wellington. Volcanic catastrophes are still very possible.

Later volcanicity hereabouts produced the island of Rangitoto which dominates Auckland Harbour; this erupted some 800 years ago, besides many other not-so-old volcanic islands and One Tree Hill on shore (now a Maori memorial) which is a perfect cone with a surrounding gully like Vesuvius. In 1953, when the Queen was visiting New Zealand, a mud flow from a broken volcanic dam near here destroyed a railway bridge and train, killing more than 100 people. So everything has not stopped, though present activity is but a shadow of what it was at the time of the eruptions of Rotorua and Taupo. The latter, for example, scattered ash over at least 10^7 km^2 with a minimum volume of 700 km^3. An ignimbrite phase yielded at least another 500–1000 km^3 (Froggatt et al. 1986). This happened about 254 000 years ago and must have been the largest eruption in the Southern Hemisphere in Late Quaternary times. One can see the ash draped over the pre-existing topography, giving the appearance of anticlines and synclines. Another eruption thereabouts, that of Mount Curl, scattered tephra as far as 10 000 km from the source and may well have caused disruption to the atmosphere and climate (Froggatt et al., op. cit.).

In the western United States everyone is well aware of the unexpected eruption of Mount St Helens on 18 May 1980, which was mostly built up only a few thousand years ago. Mount Mazama exploded 6600 years ago, depositing ash over most of the north-western states and the hole that was left is filled by Crater Lake. One of the most dramatic sights ever seen by humans must have been the lava flow which cascaded more than 900 m over the edge into the Grand Canyon less than 10 000 years ago after the arrival of the native Americans.

What is claimed to have been the Earth's greatest eruption occurred at Toba in Sumatra some 75 000 years ago (Rose & Chesner 1990). This is said to have yielded at least 2800 km^3 of rhyolitic rock, in a sheet covering 20 000 to 30 000 km^2 and a thick welded tuff. In addition at least 800 km^3 of ash was deposited over the Indian Ocean and southern Asia. The eruption must have had a fundamental effect on the contemporaneous atmosphere, with the dispersal of ash and gasses worldwide. It is said to have been two orders of magnitude larger than any historic event and must leave one doubting the competence of the present as a key to the past. Most surprising of all, it is only thought to have lasted some 9 to 14 days.

My first head of department and petrological hero, the late Professor H. H. Read, introduced the more realistic triple classification of rocks into *neptunic*, *volcanic* and *plutonic*. Considering the last named, that is the rocks formed deep below the surface such as granites, he emphasized (1949) that the emplacement of these was related to certain phases of orogenic activity and, like them, was inevitably episodic.

So there was never anything gradual or continuous about igneous activity, either volcanic or plutonic and here surely, I am entitled to use the term 'catastrophism'. In modern times it has certainly always been catastrophic for those people living in the vicinity.

12

It's the only present we've got!

I thank my colleague Geraint Owen for spontaneously and unknowingly giving me the title of this chapter, when we were discussing the misleading nature of some of today's processes.

The basic principle of uniformitarianism is, of course, that we can use the processes going on at the present time to interpret the events of the geological past. We must ask ourselves, however, whether our ephemeral present is typical of the infinite number of fleeting presents that have passed in the course of earth history. It may be that it is a very odd and atypical present that we have to use to try to understand the past.

A great deal depends on what we mean by the 'present'. When did our present 'present' start? Is the key provided by that present long enough to unlock all the difficult doors of the past? Lyell interpreted the present as meaning historic times. But when Lyell wrote his *Principles* (starting in 1830) the Pleistocene ice ages had not been recognized. The Lower Palaeozoic and certainly the Precambrian periods were virtually unknown. Much of the world had not been explored geologically. Lyell was looking at the world through the eyes of a well-educated European. One wonders if a much earlier Chinese Lyell or an Inca Lyell would have thought the same way. It is part of our unconscious racial prejudice when we say that Christopher Columbus discovered America (he wasn't the first European anyway). Of course, the native Americans knew that America was there all the time, after all they had lived there for many generations. Their various tribes knew about the Niagara Falls, the Mississippi and the Grand Canyon (in fact some of them lived in the last-named). We say that Speke discovered the source of the Nile in Lake Victoria, but the local Africans had known Lake Victoria long before the Victorian explorers and probably even before *Homo sapiens*.

As discussed in Chapter 11 many early humans must have seen geological phenomena far more violent and spectacular than any we know in historic times, including the last great volcanicity across northern Europe from the Auvergne to Romania and the explosion of Santorini which may have given rise to the Atlantis legend. In New Zealand the first Polynesian immigrants may have seen and suffered some of the last huge volcanic explosions in North Island.

Looked at the other way round, we must ask ourselves if our present is really all that typical and we must always accept the basic constraint that it may be a very odd period in which we now live.

Before the 'beginning'

Everyone knows, or at least accepts, that things were very different in the early days of earth history. There surely our strict uniformitarian approach does not apply.

I would not recognize one myself, even on a bright summer day with the wind behind me, but I am told that a distinguishing feature of the first two thousand million years or so of earth history were the komatiites. These were ultramafic lavas which just do not occur above the Archaean.

Walker maintained (1982) that there is little evidence of climatic extremes in Archaean times, notably from the evidence of glacial deposits, though evaporites are said to be no more or less abundant than in later geological periods. Geologists differ in the details, but one can say that there was a time on Earth before the first appearance of red sediments which indicate the presence of enough oxygen in the atmosphere to make oxidation possible. The early atmosphere must have had a much higher proportion of carbon dioxide than in its later history. I always talk of oxygen as the first atmospheric pollutant! For this we can blame the appearance of plants and photosynthesis, which in turn made possible the evolution of animal life. D'Argenio produced a simple graph (personal communication 1988) showing the supposed build-up of oxygen in the atmosphere and hydrosphere until the first banded ironstones appeared in many parts of the world nearly two thousand million years ago.

In a way this is all surprising in view of the theoretical probability that the sun was less bright early in earth history, though I always prefer geological facts to astronomical theory. Similarly the supposedly more rapid rotation of the Earth, with the Archaean day only lasting about 15 hours, coupled with smaller land-masses, would have produced a completely different pattern of ocean currents to what we know today. In fact world palaeogeography changed many times, and I tried to reconstruct the oceans and the currents as they were in Jurassic times (Ager 1975a).

All this, combined with a strong 'green-house effect' due to the much greater proportion of carbon dioxide in the atmosphere, means that many of our uniformitarian presumptions do not apply in Archaean times. There is also the question of the so-called gravity constant, which geophysicists tell me is immutable. I can only say that, in spite of Newton and all that: 'I have heard that story before'; I remember the time when the poor innumerate geologists were enthusiastic about continental drift, but the geophysicists said that it was impossible (now they claim all the glory). So far as the gravity constant is concerned, a variable one, for whatever reason, could mean a very different world in the past, including one that was much smaller than it is today. Though the idea is ridiculed by many or most geologists, with the notable exception of Carey (1988), I see a great deal of evidence, especially in fossil distributions, for an expanding Earth (Ager 1986c). I cannot otherwise see how you can have all the oceans widening simultaneously. Subduction alone does not solve that problem.*

The organic world has played an important part in changing the processes which affect the surface of our planet. Thus it must have been a very different place before the evolution of grass produced a close ground cover, thus reducing the rate of surface erosion and providing the food for many groups of newly evolving mammals.

The most obvious way in which our present is atypical of most of the record is that we live in the aftermath of a great series of glaciations. Glacial theory with huge ice sheets covering much of the northern hemisphere would have been anathema to Charles Lyell, since it was so 'catastrophic' in concept, though he was well aware of the importance of changes in climate and recognized the probable existence of glaciers where there are none today. In fact most of the leading geologists of the day, such as Murchison, rejected the idea.

There are several features of our present world which clearly relate to the later stages of a major glaciation. One is the generally low sea-level due to the water still being retained in the ice caps. So we have great lengths of rocky shorelines today which are rarely seen in the geological record. This was discussed in Chapter 6. The sandy beaches of the world are slowly disappearing. It has been estimated by Roland Peskoff of Lyon that 70% of our present beaches are slipping into the sea.

Another way in which today differs from yesterday is in the extent of carbonate deposition. Apart from the poor old Bahamas Banks, which have been worked to death by sedimentologists, together with a few other small areas such as Sharks' Bay in Western Australia and the south end of the Persian/Arabian Gulf, there is nothing today to compare with the vast spreads of limestones and dolomites which extended across the continents repeatedly in the past. These were discussed in Chapter 7. One might again blame this on the recent glaciation since carbonates seem to go with general global warming. Nevertheless, it has always puzzled me that the Late Precambrian Schiehallion Boulder Bed of the Scottish Highlands, which is generally accepted as a glacial deposit (and can be traced all the way up to the Varanger tillite of arctic Norway) is

* I always object to the term 'sea-floor spreading' on the grounds that it was the oceans that spread, not the seas. None of the latter, such as the Baltic, the Black Sea or even the Mediterranean show evidence of magnetic striping and spreading. The Red Sea is the obvious exception, where Girdler & Styles (1974) showed that the opening was distinctly episodic.

closely followed by carbonates. A tillite of approximately the same age has been recognized elsewhere in the world, notably in Brittany and beside the Yangtse River near the famous gorges (see Chapter 4).

Volcanicity seems to be at a very low level today compared with what it was even in the recent past. The flood basalts of the past, such as those of the Deccan Plateau in southern India or those of the Columbia and Snake River areas of the north-western USA were vastly greater than anything known in recent history. We speak of the Pacific girdle of fire – the circle of volcanoes around that ocean. But if we look at the major eruptions around the Pacific in recent times we can list: Mount Baker in north-west USA (1870), Krakatoa in Java (1883 and again in 1928–32), Mont Pelée in Martinique (1902), Lassen Peak in the USA (1917), Mount Paricutin in Mexico (1943), Mount Agung in Bali (1963), Mount St Helens in the USA (1980), Mount Chichon in Mexico (1982), Nevado del Ruiz in Colombia (1985); in Japan there was Sakurajima (1945) and Mount Unzen (1991); finally, as I write this (in July 1991) we have the news of the disastrous eruptions of Mount Pinatubo in the Philippines, which exploded after being dormant for some 600 years. If we consider the whole girdle in that way, is that continuous or is it episodic? It is difficult to judge when we look through human eyes, but for any one place, of course, these are certainly rare events.

The record of earthquakes in historic times is also interesting. Ambraseys (personal communication 1968) deduced from ancient writings and from whether or not ancient buildings were designed to resist them, that earthquake frequency since the coming of city dwellers has been distinctly episodic. Though Joshua attributed the fall of the walls of Jericho to the blowing of trumpets and the shouting of the Israelites (before they slaughtered every man, woman, child and animal in the city in the charming way of the Old Testament), it seems more likely that an earthquake was responsible. In fact further earthquakes shook Jericho in the years AD 600, 1033 and lastly on 11 July 1927. A similar cause may be blamed for the fall of Sodom and Gomorrah. All these relate to movements of the Dead Sea Rift along the line of the River Jordan. There are, of course, such areas of repeated earthquakes at plate margins, such as California and Peru, and in some areas such as Japan many minor earthquakes are recorded every day. Perhaps the most famous earthquakes were those in Lisbon in 1755, the San Francisco earthquake of 1906 and in Europe the Messina earthquake of 28 December, 1908, which is said to have killed 77 000 people. In my lifetime there have been the disasters of Quetta in Baluchistan, Tokyo in Japan, Alaska (where most of the damage was done by mud-slides) and more recently those in Iran, Armenia and Central and South America. Considerable damage and loss of life is caused by the associated tsunamis, which were discussed in Chapter 9. Disastrous landslips, triggered by earthquakes, occur repeatedly along the coast of Peru and neighbouring countries.

Earthquakes, therefore, are clearly episodic, at least in human experience, even in the most notorious earthquake areas.

Comparison with the past is more uncertain. I think of the one in Late Jurassic times on the east coast of Scotland, which sent large boulders, such as the so-called 'fallen stack' of Helmsdale, down into deeper water. I think of the Herbison Walker quarry in the Baraboo syncline of Wisconsin, where one sees great fallen blocks of Precambrian quartzite in a matrix of Cambrian sandstone filling a channel in the quartzite. This was clearly a result of landslipping, presumably triggered by an earthquake. Nearby there are clastic dykes (see Chapter 8).

In our short-sighted way, we tend to think of our present global warming as a one-off phenomenon, brought about by man's own foolishness. However, the geological record shows that it has happened many times in the past, with the spread of limestones across the continents. This happened, for example, in the Mid Silurian, the Late Devonian, the Early Carboniferous, the Mid Jurassic and the Late Cretaceous periods (Ager 1981a and herein Chapter 7).

Even extra-terrestrial impacts or near misses are known at the present day. On the morning of 30 June 1908 the Tungushka meteor is thought to have exploded 5 to 8.5 km above the ground in Siberia (60°55′ N, 101°57′ E). It left no crater, but the forest was flattened for many kilometres around and about 1000 reindeer were killed. Globules of glass were found and the iridium content of the soil was increased by a factor of four (see Chapter 13). On 30 October 1937, the Earth had a narrow escape when the asteroid *Hermes*, weighing about 500 000 tonnes passed by closer than any known before. I was a keen astronomer at that time, but no-one told me about it!

So I tried to think of something that has happened in recent times which I did not know in the past record. I was struck by the work of my colleague Sam Freeth on the Lake Nyos disaster in Cameroon on 21 August 1986. This has been described and discussed in a number of papers (Freeth & Kay 1987, Freeth 1990, Freeth *et al.* 1990). Lake Nyos is 200 m deep and fills a volcanic crater in the north-west corner of the country. Heavy rainfall and/or a cold and steady wind brought an influx of cold water into the lake, which is normally at a temperature of about 20°C. The warm water naturally rose from below triggering the release of toxic gasses held under hydrostatic pressure. About 99% of these gasses was carbon dioxide. An aerosol of water and gas swept down the valleys to the north and sent huge waves, up to 75 m high across the lake, flattening the vegetation. About 1700 people died in this disaster, together with thousands of livestock. One of the reasons for the high loss of life was that a new road had been built down the main valley along which villages grew up. It has been estimated that the disaster is likely to be repeated in cycles of less than 30 years. Chevrier (1990) reported on a visit to Lake Nyos in December 1986 and January 1987, a few months after the catastrophe. He noted similar events taking place, but of a much weaker intensity. As the whole area had been evacuated, only five people witnessed these later happenings. It has also been suggested that the wall of the crater, which is slowly being eroded, will give way and a flood of water will rush down the valley. So any local inhabitants remaining will have the alternatives (though not the choice) of being drowned or asphyxiated.

So here, I thought, was something happening at the present day of which I knew no equivalent in the past. When I asked Sam Freeth he suggested that such an occurrence might be recognized in the geological record by a deposit of limonite and a concentration of fish and of birds (personal communication 1991). Concentrations of fossil fish are not uncommon, but concentrations of fossil birds and other flying animals are extremely rare. I thought immediately of Lake Avernus, near Naples, which was the classical entrance to the underworld and where birds flying across it are said to fall dead because of the noxious gasses being emitted, as from Lake Nyos. Near here is the Grotto del Cane, where carbon dioxide kills dogs near the ground, but not humans who are higher up (Ager 1955b). A thick bloom of limonite (hydrated iron oxide) would result from the upwelling of bottom waters rich in iron bicarbonate. The toxic gasses would kill all the life in the lake and birds flying over it.

Concentrations of fossil fish in lake deposits are found, for example, in the 'Old Red Sandstone' (Devonian) of Scotland and elsewhere in Europe. They are also found in fresh water diatomite deposits, for example in the Miocene of Sicily and the Oligocene of Romania. Fossil birds, on the other hand, and other flying animals are very rare because of their mode of life and the fragility of their skeletons. Franzen (1985, Franzen et al. 1982) suggested the rapid escape of noxious gas to explain the birds, bats and insects which are unusually abundant in Eocene oil shales in the Grube Messel in Germany. Going back before the times of abundant birds, there are abundant pterosaurs (flying reptiles) accompanying mass mortalities of fish in organic-rich sediments in the Mid Cretaceous Basin of Chapado do Araripe in north-east Brazil (Dave Martill in letter to Freeth 1990 and personal communication 1991). So it is apparent that even such unusual events at the present day may be recorded in the past.

I have already commented on the unusual proportion of rocky shorelines (with abundant life) which characterize our fleeting present, but not usually the past. I never cease to be astonished at the widespread occurrence of pebble beaches in north-west Europe, almost wholly composed of flint pebbles from the Upper Cretaceous Chalk. The mind boggles at the amount of Chalk which must have been eroded to produce them. I know of nothing comparable in the geological record, apart from a very recent deposit in the Boulonnais of northern France, quaintly (and quite wrongly) called 'Pseudo-Aptien' (Ager & Wallace 1966a).

Even some of the supposedly 'continuous' processes going on today may, in fact, be quite 'sudden' in geological terms. Thus it has recently been shown that the Great Barrier Reef, off the east coast of Australia, is much younger than was previously supposed, which may contradict what I said in Chapter 7 and indicate that another spread of carbonates is on the way. Similarly, subsidence may be quite sudden, as in the island of Gozo, Malta, where there are numerous examples of circular masses subsiding into solution cavities in the underlying Miocene limestone (Figure 12.1). The village of Maqluba (Maltese for 'overturned') suddenly disappeared into the ground in 1343 leaving only the church on the rim of the collapse structure (Zammit-Maempel personal communication 1974). Naturally this was blamed on the wickedness of the inhabitants.

Figure 12.1   One side of the circular collapse depression of Qawra, island of Gozo, Malta. Photo DVA.

In South Wales, cars occasionally sink into holes in the roads over old coal workings and in north-west England there is subsidence due to the old salt workings. That brings us naturally to the species which is blamed for so many of the misfortunes that have afflicted the world in recent years. I refer, of course, to the works of that newcomer *Homo sapiens*. Besides the many bad things that have been done I should mention one of the concerns which has never bothered other species – the fate of organisms on our planet. This is the understandable concern of the conservationist for 'endangered species' such as the whales and the elephants. But this is merely a continuation of the long history of life on the Earth, with extinctions followed by the emergence of new forms. I never saw an Indian elephant in more than a year in that sub-continent, but I am particularly fond of the African species, which I have seen in large numbers. The massacres by the ivory poachers must obviously be condemned and prohibited. We try to forget now that elephant hunting was a favourite sport of white settlers in the past, with the prospect of financial gain from the ivory (as one learns from Wilbur Smith's novel *When the Lion Feeds*). I was surprised in East Africa at the number of elephants with only one tusk and wondered if this was a form of natural selection in reverse, since such deprived beasts must be less attractive to the poachers. But looking at the present as long-sighted geologists should we not expect extinctions?

Of course humans have had a considerable effect on their world, more than any other species except perhaps for the earthworm and grass. Self-satisfied people in the developed countries criticize those in the poorer parts of the world for cutting down their tropical forests, forgetting that the same thing happened at an earlier period when most of the temperate forests of the northern hemisphere were destroyed. Believe it or not, I just took a short break from my processor and idly turned on the television to see the catastrophe wrought on the area around the Aral Sea in the Commonwealth of Independent States. The sea has shrunk and become highly polluted by the chemical fertilizers used in the past to encourage the local coffee crop. Of the ten thousand fishermen, there are none left because there are no fish. Former fishing villages are now in the desert and people have to go many kilometres to fetch drinkable water. A Tanzanian student I persuaded to talk to a local club was proud of the fact that no-one in his country (usually meaning, of course, no woman) now had to walk more than four miles (about 6.5 km) for water!

Deserts are spreading all over the world. The Sahara is quite rapidly moving south. My son spent three years in Burkina Faso, West Africa, building dams to form reservoirs for villagers who were desperately short of water for their crops, their livestock and themselves. He told stories of women sleeping by the wells at night in the hope of there being a little dribble of water at the bottom next morning.

The cutting down of forests has caused disastrous soil erosion, notably in northern India. I saw a great area near Ducktown, Tennessee, eroded and devoid of vegetation thanks to the yellow fumes still pouring from the copper smelting works in the background. In a picture I have of the Swansea Valley in the last century, one can hardly see the valley at all because of the smoke pouring from the many works when this was the 'metallurgical centre of the world'. Unfortunately people think it is still like that and surrounded by coal tips, whereas in fact it is a very beautiful place. The great steel works at Port Talbot across the bay still does harm, however, as one can see from the dead trees on the escarpment opposite and the resultant land-slipping. Much of the troubles of mankind have been blamed on over-population and over-grazing by his livestock, but the greed of the 'western' nations has played a big part in it.

In other places, notably the Netherlands, the world has been changed by the reclamation of great areas of land from the sea and we have drained large areas such as the fenlands of eastern England. Conversely valleys have been flooded to create great reservoirs, as in northern Spain, where drowned villages emerge again at the end of the summer.

Concrete and tarmac have covered much of the countryside in cities all over the world, even in the most unlikely places. Wellington, New Zealand, was planned on a grid-iron pattern in London before anyone told them that there was a mountain range running through the site. Japanese cities have grown at an almost unbelievable rate, so that this visitor's impression of that country was of bursting cities, paddy fields on even the smallest flat areas of countryside and practically unused mountains. I thought these last would be ideal for sheep, but the only sheep I saw in Japan was one stuffed in a museum in Hokkaido. At least they do grow grapes on them in a few places to produce some very good wines. I liked particularly *Chateau Lion, 'mit en bouteille dans nos caves en Yamanashi'* as it said on the label.

So man has changed his world and made it less suitable for interpreting the past, though I always argue that we are just one more species 'doing its thing' like every other species before us. Every species today, and presumably in the past, has pursued a policy of 'my species right or wrong'. The survival of the species is more important than anything else. We are just more efficient at changing things and at killing other organisms, including our own kind. I say of the cannibals that they are no more fond or efficient at killing each other than we are, but at least they got some nutritional value out of it. Therefore all human activity, going back to prehistoric times when the first forests were cut down, has always been, or intended to be, to the species' own advantage.

I read once that the function of the lights on deep ocean fish was to enable predators to catch them. This must be nonsense as are the suggestions that bryozoans spent their lives 'building their own tombs' until they could no longer feed or that oysters rolled themselves up until they could no longer open their valves. In spite of what I said earlier about 'mistakes' in evolution, I cannot believe that any species has ever evolved naturally to its own disadvantage. Where it seems to have done so, it has usually been because of human interference, as with the pigeons and with the livestock that cannot give birth without human help. What we see today is that every species struggles for its own existence and the perpetuation of its selfish genes.

If mankind is causing the extinction of many species today, then I cannot see that this is so very different from what happened in the past, albeit at a slower rate. The ammonites and belemnites may have become extinct because of the evolution of new predatory fish and mammals (Ager 1976). Did the brachiopods become less important at the end of the Palaeozoic era because the bivalve molluscs took over so many of their habitats? No doubt we are living at a time when organic changes are taking place very rapidly with the extinction of many species due largely to the action of one species, *Homo sapiens*.

If we consider the landscape, this seems to be unusually placid and dry at the moment (though not so much so as in the Triassic). Many want to change it more rapidly, others to return it to what it once was. So far as these islands are concerned and looking at it as an Englishman, I am something of a romanticist and yearn for the winding country lanes, small corn-stooked fields and flowery hedgerows of my youth. But I am a little tired of people who imagine a past golden age, a sort of Arcady that we have lost. Such an Arcady never existed. There was nothing 'natural' about thatched cottages, small fields and winding lanes. An Anglo-Saxon conservationist, if there ever had been such a person, would be horrified at what we have done to the strip farming and the forests full of wild boars and wolves. We always forget, when talking of the Industrial Revolution, that people flocked to the towns because the crowded slums and soulless work in factories seemed better than the squalid cottages and back-breaking work in the countryside.

Even in our brief existence on this planet, however, we have really left little for the future to remember us by. Most of our works will soon, in geological terms, have disappeared. T. S. Eliot wrote of mankind:

Their only monument the asphalt road
And a thousand lost golf balls.

Perhaps if he had written that in a later decade he would have included the imperishable plastic bottle. It is not much of a record for a conceited species to be proud of and, with the observations of Copernicus and Darwin, it puts us in a very humble place in earth history. No doubt, assuming the usual duration of a species, when geologists come from other planets to examine our planet in a few million years time, they will find more stone axes than the artefacts of the fleeting culture and technology which served us only in the dying moments of our species.

I sometimes think of the fate of the Viking colonists in Greenland, who lingered on for hundreds of years till the fifteenth century. By then they were completely out of touch with the rest of mankind and the last survivors must have thought that they were the last people on Earth. Imagine being the very last man or woman! The past can be a guide to the present, rather than the other way round.

There is a Japanese saying: 'Visit the past to know the present'.

13
Jupiter or Pluto?

I doubt if Shakespeare knew much about geological theories, otherwise he might have written:

The geologist's 'eye, in a fine frenzy rolling
Doth glance from heaven to Earth, from Earth to heaven'.

So, in conclusion, we must discuss whether it was all due to Jupiter hurling down his heavenly thunderbolts from above or to Pluto stoking his hellish fires down below.

In recent years it has become increasingly fashionable among geologists and others to blame many of the dreadful things that have happened to this Earth of ours on the impact of extra-terrestrial bodies or on other astronomical phenomena. Alternatively many geologists, perhaps the more conservative among us, have preferred to think of internal processes within the Earth such as convection currents below the crust, mantle plumes and plate tectonics, as the moving forces which were responsible for most of the phenomena which I have discussed in this book. I shall try to be fair to both schools, but although the former sounds more 'catastrophic' and therefore more in line with my philosophy, I must declare myself more in favour of Pluto for many of the catastrophic events that have been inflicted on our planet.

First then Jupiter: though thought of as a very modern idea, blaming earthly catastrophic events on the heavens is a very ancient way of explaining disasters. Thus in the Bible, in that strange section called Revelations (Chapter 8, verses 7–11), it is said that seven angels sounded trumpets. The first invoked hail and fire 'which burned a third of the trees and all the grass'. At the second trumpet call 'a great mountain burning with fire was cast into the sea . . . ; And the third part of the creatures which were in the sea . . . died'. And when the third angel sounded his/her trumpet 'there fell a great star from heaven, burning as it were a lamp . . . and many men died'. I must say that I have never been very fond of angels as I could never understand their pectoral girdles, with two arms and two wings! Anyway, it would seem that holy writ covers itself for both Jupiter and Pluto. Whether one takes this seriously or not, there still remains the possibility that it recalls something in distant folk memory.

Turning to a much later record, according to Shakespeare, Calpurnia (Caesar's wife) said: 'When beggars die, there are no comets seen; The heavens themselves blaze forth the death of princes', though I doubt if this correlation could be proved statistically. Of course another popular culprit for disasters was man's immorality. This goes back to Sodom and Gomorrah in Genesis, destroyed with brimstone and fire. Even an exceptionally severe winter, round about 400 BC, was blamed by the Greeks on the impiety of the people. More recently there was the sudden collapse of the Maltese village of Maqluba in 1343, which I mentioned in the last chapter.

Way back in the seventeenth century, the English inventor Robert Hooke carried out experiments to test the possible causes of the craters he saw on the Moon through an early telescope. Already he was considering the possibility of their being produced by the impact of foreign bodies and he produced very similar structures by throwing heavy objects into thick mud. He did not think this possible, however, so he tried boiling alabaster and produced the same result. He therefore decided on a volcanic origin for the craters on the Moon (Drake & Herman 1988).

When I was very young (a long time ago) I remember arguing with my oldest friend about the same alternatives, though I cannot remember which of us took which side. Modern Jovian thought has concentrated on the idea of the arrival on Earth of extra-terrestrial bodies, variously called comets, asteroids and meteorites. From what we know now of impact craters on the 'back' as well as the 'front' of the Moon, on the satellites of other planets, on Mercury and even on Venus with her thick atmosphere, it is obvious that such impacts have been one of the main physical processes affecting our Solar System. It is also obvious that the Earth must have suffered in the same way as her cosmic neighbours (Ager 1988a).

Until quite recently the debate continued. Thus von Bulow (1960) compared the crusts of the Earth, the Moon and Mars. He recognized, among other things, the probability of subcrustal currents on all three. He also argued briefly that the lunar craters were volcanic in origin and opposed an exclusively meteoritic explanation. However, the meteoritic school now seems to have won on most counts, although volcanoes have been found on several bodies in the Solar System.

It must be remembered, however, that intense meteoritic bombardment of the Moon tailed off some four thousand million years ago and all that we can hope to see on Earth is the very end of this episode. It may therefore be significant that most of the impact structures now recognized on Earth are in the ancient shields. This applies for example to the elliptical structure at Sudbury in northern Ontario which contains the richest nickel deposit in the world. It is tempting to associate the nickel with the nickel-iron meteorite which was probably responsible, but I am told by the more down-to-earth mining geologists that the impact may have simply released a flow of nickel and other ores into the crater, as a result of the melting of surrounding basic and ultrabasic rocks. The evidence for the impact shown me by my former research student Paul Copper is quite convincing with shatter cones and impact breccias (Figure 13.1).

Figure 13.1 Shatter cone (with associated impact breccia) in the McKin Quartzite, Ramsay Lake near Sudbury, Ontario, Canada. My friend Paul Copper as scale. Photo DVA.

I should in this chapter again mention the 1908 Tungushka meteorite in Siberia, which I discussed in the last chapter. This at least shows us that these things do still happen at the present day. I have not visited Siberia (though I have flown over it), but I have crossed a paradoxically snow-covered desert to see Meteor or Barringer Crater in northern Arizona. This arrived only yesterday, geologically speaking, and the meteorite buried itself in Permian strata about 50 000 years ago, dated by a small lake deposit on its floor. It is about 1200 m across and nearly 170 m deep. Nickel-iron was recovered in borings, but Barringer's hoped-for fortune did not materialize because he drilled in the centre of the crater, not realizing that the meteorite must have struck obliquely and what remained of it was probably buried under the northern rim (Figure 13.2). Again it is a matter of preservation, since this crater has only survived as a topographic feature because it is in a desert area with a very low rainfall.

Figure 13.2   Meteor Crater, northern Arizona, USA from the air. Bought in a shop
and source unknown.

In Europe we have the Reis and Steinheimer craters in south-west
Germany, with Miocene lake deposits. It has always struck me as odd
that they should be near a cluster of Tertiary volcanic craters, but the
latter differ in age and this cannot be more than a coincidence (Ager
1980, Figure 8.12, p. 196). The largest ancient impact crater in Europe
is Siljan in central Sweden, which even shows up as a circular structure
on a postcard size geological map of that peaceful country. If one goes
up the aerial ropeway on the south rim, one has a fine view of the crater
('the finest view of Siljan in the world' they say on a notice here, which
is hardly surprising). One sees clearly the circular lake which fills the
edge of the crater (Figure 13.3). The impact is said to have affected
Ordovician rocks here lying on the ancient Fenno-Scandian shield.

   There are about 100 to 120 impact structures or 'astroblemes' so far
recognized on the Earth, though many of these are disputed. The largest
of all was thought to be the Manicouagan feature in Quebec (also on
the Canadian Shield). This is some 65 km across. I was told of an even
bigger one, 70 km across, recently discovered by Wu Siben on the
borders of Hebei Province and Inner Mongolia in China. This was called
the Duolum Crater. The news came to me on a Christmas/New Year
card from my friend Liu Benpei in 1987. This crater could be dated
quite accurately, geologically speaking, since the extra-terrestrial body

Figure 13.3   Lake Siljan filling the rim of an impact crater in central Sweden. Photo DVA.

had struck Upper Jurassic andesites and it was overlain at its margins by Lower Cretaceous coal-bearing sediments. I duly passed on the news to Richard Fifield of *New Scientist* who wrote a sensible article about it, mentioning that such an impact had been blamed for the extinction of the dinosaurs. This titbit was immediately seized upon by the popular press (i.e. the *Daily Mail*, 23 February 1987) suggesting that the Duolum impact was responsible for the extinction of the dinosaurs, ignoring a time gap of some 80 million years! That illustrates the nonsense that so often creeps into this debate.

The height of such idiocy is illustrated by the cult of Velikovsky, who postulated major collisions up into historic times. I was repeatedly urged to read his works by one of his adherents. Whilst giving a course of lectures in Amsterdam and Leiden and left in the evenings with nothing to read, I chanced upon a copy of one of his works in a back-street bookshop. I will not encourage such pseudo-science by giving a reference. I will be presumptious enough, however, to quote my own subsequent letter to my correspondent about the book I had read:

I was frankly appalled. It was far worse than I had been led to suppose. I can assure you that 90% of it is the most unmitigated nonsense. He does not present any geological data whatever to support his views. He merely cites very out-of-date authorities (some 18th century!) in a highly selective manner. It is rather like someone quoting mediaeval alchemists to disprove modern atomic theory. He puts together observations which have no possible connection with each other. He appears to have no concept whatever of the geological time-scale and associates things which happened tens of millions or even hundreds of millions of years apart. He seems oblivious of the hundreds of thousands of successive fossiliferous horizons there are in the rocks of almost every part of the earth's surface and even seems to be unaware of the normal mortality of living organisms. The idiocy of some of his dogmatic pronouncements is illustrated, for example, in the [alleged] transport of vast numbers of large vertebrates from the tropics to the arctic by a great wave that did not apparently carry with it a single marine organism.

I am sorry if I appear to be neurotic about this, especially as Velikovsky seems to be on the side of the catastrophists, but I do not want to be associated in any way with such nonsense. This, together with the writings of the Californian 'creationists' are the reason for my disclaimer at the beginning of this book.

We come then to scientific views of extra-terrestrial impacts. It has often been suggested that such events were the cause of mass extinctions of life on Earth. De Laubenfels (1956) and Digby McLaren (1970), long before the present fashion and before most people had heard of iridium, suggested that the arrival of a large foreign body in the Pacific may have caused a surge of turbid water which wiped out the sessile suspension feeders on the world's continental shelves at the end of the Frasnian Age of the Late Devonian. This is a well-documented mass extinction and one of the few that I really believe in. On the other hand, Paul Copper suggested (1986) that this extinction was caused by global cooling, due to a change in ocean currents with the coming together of Eurasia and Gondwana.

Urey (1973) used the presence of tektites at the base of the Tertiary as evidence of a cometary collision at that time, which heated up the atmosphere and killed off all land plants and animals. This would account for the extinction of the over-publicized dinosaurs, but hardly for the major marine extinctions at this time and the big changes in the floral world had happened much earlier, in Early Cretaceous times.

Victor Clube suggested (1978) that our galaxy had a violent history and later he and Bill Napier argued for a theory of terrestrial catastrophism which really started the recent interest in asteroid impacts on the Earth (Napier & Clube 1979). Later they produced a remarkable book (1982) which was subtitled *A catastrophist view of Earth History* presenting a convincing argument for a giant comet that terrorized mankind in prehistoric times. They put together a great deal of evidence from a variety of sources that speaks of a violent past and a hazardous future.

However, the paper that really accelerated the idea of impacts causing mass extinctions was that by Alvarez, father and son, with Asaro and Michel (1980). These reported high levels of iridium at the Cretaceous–Tertiary junction at Bottaccione near Gubbio in the Italian Apennines. Iridium is a member of the platinum group of elements and is particularly characteristic of meteorites. Though it was only present in minute quantities (parts per thousand million or trillion) this unusual peak (or 'spike' if you are an American) was claimed as evidence of a major extra-terrestrial impact at the end of Cretaceous times. It was suggested that the Earth passed through a long episode of darkness which killed the plants and therefore the herbivorous dinosaurs and, in due course, the carnivorous dinosaurs. The flying pterosaurs went too. Probably more significant and more difficult to explain, was the complete collapse at this time of the whole marine ecosystem from the minute coccoliths via the ammonites and belemnites to the large marine reptiles. It has been suggested that the marine extinctions too could have the same cause, perhaps by the acidification of the oceans. Ken Hsü has commented on the presence of cyanide in comets (Hsu 1986), which might be a better explanation for the marine extinctions, but I would expect to find some evidence for either of these chemical causes on the Earth.

To be pedantic, the very title of one of the Alvarez papers reveals a fundamental lack of geological understanding. The title reads 'The end of the Cretaceous: Sharp boundary or gradual transition?'. If they had said 'top' it would have made sense, but 'end' implies time and one cannot emphasize too often that there cannot be gaps in time. All time boundaries are, in a sense, both sharp and gradual. The question is whether there was a halt in deposition at the critical localities where high iridium anomalies have been detected. The answer is almost certainly 'yes'. Breaks in deposition are the rule rather than the exception in the stratigraphical record as I emphasized in my previous book on the subject (1981c, Chapter 3). I would suggest that the iridium boys should look for their minute anomalies at known breaks in deposition, such as the layer of bored phosphatic nodules at Gorodishchi on the Volga (*op. cit.* Plate 3.4) which seems to be the equivalent of the whole of the Portlandian Stage that marks the top of the Jurassic in Britain (Casey 1968). Layers of phosphatic nodules on the present sea floor, for example on the Chatham Rise, east of New Zealand would surely merit such geochemical analysis, since they seem to represent a time interval of many million years (Cullen 1980).

Hsü has argued (1986, Chapter 3) that on the basis of the foraminifers there was 'No Chasm at Gubbio', i.e. no great gap, but Finn Surlyk (1980) drew attention to Wezel's detailed work in which he found much reworking of the microfossils and therefore of gaps in the classic sequence. As Surlyk says 'Wezel's paper is so detailed that the burden of proof now rests on his opponents'. With colleagues he also found (Wezel *et al.* 1981) several layers rich in iridium in the Apennine succession. Wezel also recorded (1981) an iridium anomaly about twice that at the boundary, some 240 m lower down.

Alternatively, Tony Hallam (1989) has suggested that major marine regressions, as at the end of the Cretaceous, were enough to explain the mass extinction of marine faunas. Obviously transgressions would be of advantage to marine life, in providing them with more territory and more ecological niches. Conversely, regressions would have had the same effect on land life (Ager 1976, 1981a). Similarly Jaeger (1986) has argued that the Late Cretaceous transgression resulted in an unusually equable warm climate and optimum conditions for organic evolution, but the succeeding major regression had a profound effect on the world's life. He saw no need to postulate the impact hypothesis for the major extinctions at the end of the Cretaceous.

As I have mentioned before, Hallam has argued (1984) that the extinctions caused the iridium rather than the other way round. In other words, the extinction of nearly all the coccoliths and many of the pelagic forams which form more than 99% of the Chalk at the end of the Cretaceous caused a halt in the build-up of carbonates and resulted in a long break in deposition which allowed the meteoritic dust (mentioned in Chapter 8) to accumulate. There have been subsequent arguments about the possible length of such a gap, but it is very difficult to be dogmatic about such matters.

Crocket and others (1988) also recorded the distribution of noble metals in the Gubbio area. They measured the iridium peak in the boundary shales as 63 times that in the surrounding rocks. It must be said, however, that one must compare like with like. It is obvious that slowly deposited shales are likely to accumulate the rare elements at a much greater rate than more rapidly deposited limestones. Similar peaks were found at the junction in the classic section at Stevns Klint on the east coast of Denmark, south of Copenhagen. Again this was in a thin clay seam within a calcareous succession (Figure 13.4). This seam has long been known for what was taken as volcanic material and we know that there was Mesozoic volcanism just across the narrow straits in southern Sweden (Rosenkrantz & Rasmussen 1960). What is more, the Mø Clay in northern Denmark, of Eocene age, contains many bands of volcanic ash (Pedersen et al. 1975) and the Palaeocene Fur Formation of the same area contains no less than 179 ash layers (Pedersen & Surlyk 1983). Since then we also have the evidence of Cretaceous volcanicity in the North Sea. I will show later that iridium can come from volcanoes as well as from meteorites.

I have not been to Gubbio, but I have been to Stevns Klint and what I saw there was not exactly what I read in the literature (Figure 13.4), I saw typical white Maastrichtian (uppermost Cretaceous) Chalk with the usual black flints and characteristic fossils such as *Echinocorys* and *Tylocidaris* (together with other fossils such as bryozoans). There were

Figure 13.4   The junction between the Cretaceous and Tertiary at Stevns Klint, south of Copenhagen, Denmark. The hammer marks the 'Fish Clay' with its iridium anomaly. Typical flint concretions are seen in the Maastrichtian white chalk below the obvious unconformity. Photo DVA.

no ammonites or belemnites (and certainly no dinosaurs!). Above the clay band and a clear break I saw white Danian chalk with flints and the same familiar fossils as before. This is illustrated in a specimen I collected there (Figure 13.5). Marianne Bagge-Johansen (1988, 1989a) has emphasized the monotonous continuity of the Chalk through from the Cretaceous into the Tertiary. I was not so sure about the brachiopods, and blamed many of their supposed mass extinctions on the use of 'data bases' which falsely compartmentalize information (Ager 1988b). I also noted that though they may have seemed to dwindle suddenly to next to nothing in Europe at the end of the Cretaceous, this was when they really started in earnest in the Pacific area. I was forced to rethink, however, by the detailed work of Marianne Bagge-Johansen in Denmark. She made a very thorough study of the brachiopods (including micro-morphs) at the Cretaceous-Tertiary boundary in north-west Europe and demonstrated a very marked change (1988, 1989a & b). In an earlier paper with Finn Surlyk (1984) she showed that most of the extinctions were at a specific level in Denmark. They record 26 extinc-tions of species at about this level, but only eight extinctions of genera (and this may be local). For years there were arguments among palaeon-

Figure 13.5   Specimen collected by the author from the Danian at Gubbio, showing typical white chalk with black flint and two typically Cretaceous Echinoids, a large specimen of *Echinocorys scutatus* and a small *Tylocidaris* below it to the right. Width of specimen about 13 cm. Kindly photographed by Mrs K. Bryant, by arrangement with the National Museum of Wales (where the specimen, Reg. No. 92.16G.1, is preserved).

tologists as to whether the Danian belonged in the Cretaceous or in the Tertiary. The mere fact that there were such arguments suggests that the differences were not all that obvious. Generally speaking the macropalaeontologists favoured the former and the micropalaeontologists voted for the latter. The tiny fossils won because certainly there was a major change in the planktonic foraminiferans and the coccoliths at this level.

An iridium peak was also found at the same level in Woodside Creek in New Zealand, again in a boundary shale. This was confirmed by Brooks and others (1984) who recorded that the iridium value at the New Zealand boundary was higher than in non-boundary shales anywhere else in the world, including those of adjacent shale bands, *when the data were expressed on a carbonate-free basis* [my italics]. In this case I was almost convinced because of the plesiosaur less than a metre below the boundary along the Waipara River in New Zealand (Figure 13.6). This important section was described in terms of sedimentology and micropalaeontology by Strong (1984). More recently, well-dated glass

Figure 13.6   Dr Alexa Cameron pointing to the place in the section on Waipara River, near Christchurch, New Zealand, from which a plesiosaur skeleton was removed. This is less than a metre below the Cretaceous/Tertiary junction. Photo DVA.

beads or tektites have been found at this level in Haiti and (perhaps significantly) in volcanic ash in Montana. Alvarez made a fierce personal attack on C. B. Officer in the New York Times of 19 February 1988, for claiming that he had found similar beads above and below the vital junction. This indicates the strength of feeling that has been aroused in these discussions.

Iridium peaks, together with those of other elements such as gold, copper, cobalt, osmium, chromium, nickel, arsenic and molybdenum are said to have been found at more than 70 Cretaceous–Tertiary boundary sites around the world, so it is obvious that whatever produced the anomalies must have affected the whole planet. It must be said, however, that some of these elements are more reminiscent of the Earth than of space and Larry Gough of the US Geological Survey has claimed that the anomalies could be derived from the Earth's mantle. It has also been argued that iridium can be produced in volcanic eruptions. Zoller and others (1983) drew attention to the high iridium levels in the airborne particles from the eruption of Kilauea on Hawaii in January 1983. It should be noted that all the classic iridium localities known to me are in areas of contemporaneous or near contemporaneous volcanicity. It has been suggested (e.g. by Kerr 1991) that the two might be connected and that a huge meteorite might have triggered off the eruption of something like 2 million cubic km of lava in the Deccan Traps of India, exactly 64.68 (± 0.12) million years ago. This has been denied by Mitchell & Widdowson (1991), who recognized a diachronism in the basalts from north to south, either caused by a hot spot moving south under India or India moving north over a hot spot. They found 'it unnecessary to invoke a cataclysmic impact event as an explanation for the Deccan eruptions.'

So I am not completely convinced about that foreign body at the end of the Cretaceous period, though I am prepared to be open minded about it.

However, the emphasis in the literature, on the television and in the press is always on the dinosaurs, because they figure so much in the popular imagination. When I tell normal people that I am a geologist, they either mistakenly tell me about the Roman remains in their home town or they ask me for my views about the extinction of the dinosaurs. Always it comes back to the extinction of the dinosaurs. I must admit to being a little tired of those stupid great beasts, though they are the best recruiting sergeants for our subject among young people (including myself). Their importance, in my view, is grossly exaggerated. We know only too well at the present day that large vertebrates such as elephants, rhinoceroses and whales are particularly vulnerable and in danger of extinction. The popular attitude seems to expect geologists to be constantly tripping over dinosaur remains, but they are in fact very rare except at certain famous localities, such as Drumheller in Alberta and the Dinosaur National Monument in Wyoming. However, in more than 40 years working on rocks of the 'dinosaur age' (i.e. the Mesozoic) I have only encountered them twice outside museums. Once there was a known locality for small rolled bones in north-east Mexico; the other was some footprints two research students and I were led to by Berber boys in the central High Atlas of Morocco.

It should be remembered that the dinosaurs were very much in decline before the end of the Cretaceous. I wish people would not always talk about the extinction of the dinosaurs as if the whole rich diversity of those beasts through Mesozoic times all disappeared together at the end of the Cretaceous. In fact there were probably something like 350 genera of dinosaurs altogether, but only a handful of species by the end of the Era. Lewin emphasized how few species there were left at the end (1990). When I pointed this out to Ken Hsü, he commented 'a bomb on an old people's home is still a disaster' and I must agree with him. Alan Charig of the Natural History Museum in London has now listed more than a hundred suggested theories about the cause of their extinction (personal communication 1991). The latest hilarious theory I have heard is that they were too big to enter Noah's ark! More seriously, though almost equally hilarious, are my favourite theories, that they died of sexual frustration or of chronic constipation.

The first might have been caused by the perverse reptilian habit of only producing embryos of one sex, dependent upon the contemporary climate and certainly there seems to have been a marked climatic deterioration at the end of Cretaceous times. This was claimed in New Zealand, for example, by Clayton & Stevens (1968).

The chronic constipation may have been the result of the evolutionary explosion of the angiosperms (the flowering plants) in earlier Cretaceous times at the expense of the ferns, 'cycads' etc., with a resultant change from dominance by the resinous ferns and gymnosperms to the less laxative properties of the flowering plants. This may well have upset the dinosaurs' digestion. It is surely significant that almost all the living reptiles are carnivores and insectivores. The only exceptions being the herbivorous tortoises and turtles, which can stomach the alkaloid-bearing plants which the mammals cannot tolerate (Ager 1976).

It may also be significant that some of the last dinosaurs, notably *Triceratops* (which is almost the zone fossil of the topmost Cretaceous non-marine deposits) had jaws like garden shears, very well adapted to dealing with the stiff leaves of the cycad-like plants which were then on the way out (Ostrom 1964). They were clearly not suited to the softer foliage of the angiosperms. It may not have been the intention of the artist, but I like a picture I have from *Life* magazine of a *Triceratops* looking askance at an early *Magnolia* in full flower (see also Swain 1974). Another likely possibility is that some of the dinosaur families, presumably the smaller ones with bird-like pelvic girdles, took to the air and became birds.

A common theory is that dinosaurs, being reptiles, were cold-blooded and were therefore limited in their distribution, like modern reptiles, to the warmer climates and were replaced by the warm-blooded mammals coming down from the polar regions. However, many specialists now believe that the dinosaurs themselves were warm-blooded and the discovery of their remains on the north slope in Alaska, even then at a very high latitude, suggests that they were not all that intolerant of low temperatures. One problem always remains of why the mammals remained in such a minor rôle for so long, that is through the greater part of the Mesozoic. Perhaps this too was a matter of their digestive systems. It could be that the Mesozoic plants did not suit them. Years ago in a dentist's waiting room, I read a woman's magazine (which is all that is usually available in such places). I forget the romantic aspect of the story, but it told of an Australian rancher who went out driving a crowbar into the cycads on his land and pouring in creosote to kill them, as they were poisonous for his stock. Chamberlain (1965) mentioned this problem in his book on the cycads and told how the Australian government was trying to exterminate cycads with arsenic. It is well known in Britain that the common fern called bracken (*Pteris aquilina*) is poisonous for cattle. Similarly the poisonous yew (*Taxus baccata*) was planted in churchyards, where the trees could do no harm to cows and horses, but were needed to provide the wood for our major weapon in the Middle Ages, the longbow, which won several of our major (and senseless) victories, such as Agincourt. As a result many English church-yards have ancient yews.

Much more recently I read an article about Lou Gehrig's disease (named after a famous American baseball player who died of it). It is also known as ALS (amyotrophic lateral sclerosis) and is a disease which destroys the brain's motor neurons (Davies 1991). It also killed the film actor David Niven. Recent studies have concentrated on molecular genetics, but what seems highly significant to me is that the disease is particularly prevalent on the Pacific island of Guam. There the blame was put on the local habit of making flour from the seeds of cycads, which are known to be toxic. The disease was particularly widespread on that island during the Second World War when food, especially rice, was in short supply and the locals depended more on the cycad flour known as *fadang*. Another interesting titbit of information I learned recently from a Christmas present – Michael Crichton's unusual novel *Jurassic Park* (1991, p. 87) is that a common fern in the Jurassic, *Serenna veriformis*, only survives today in the wetlands of Brazil and Colombia and is very poisonous for mammals, even to mere touch. So perhaps the cycads and/or the cycad-like plants, together with the ferns and conifers, all of which flourished in the Mesozoic, were particularly poisonous for mammals and the latter did not come into their own until these were largely replaced by the flowering plants which also, perhaps, caused the demise of the dinosaurs and gave the mammals their great ecological opportunity. It is worth noting here that the angiosperms or flowering plants seem to have passed through from the Mesozoic without major change. Thus Taylor (1991) showed that the floras of the Andes persisted from the Cretaceous to the Pliocene and their changes were related to climate, altitude and palaeogeographical connections.

There were other, seemingly humble organisms, however, which in my view, were far more important than those big lumbering dinosaurs and which almost certainly passed through the dreadful events at the end of the Cretaceous unscathed. I am thinking particularly of earthworms and grass. Darwin's last book (1881) was devoted to the importance of earthworms in the formation of soil. As Pemberton and Frey reminded us recently (1990), his last words in that book were: 'The plough is one of the most ancient and valuable of man's inventions; but long before he existed the land was in fact regularly ploughed by earthworms. It may be doubted whether there are many other animals which have played so important a part in the history of the world, as have these lowly organized creatures'. Unfortunately we do not have a very good record, as yet, of non-marine burrows, but they do exist and I have no doubt that they were as abundant as those of their marine cousins. Harry Doust (1968) recorded burrows which he called *Novisia* in the terrestrial facies of the Miocene Marada Formation in eastern Libya. They are associated with casts of rootlets and may well have been formed on a flood-plain or levees. Unfortunately we are faced as always with the problem of preservation and the dry land is not a good place to live if you want to be a fossil!

Turning to grass, as Saint Peter reminds us in the Bible, 'All flesh is grass'. It feeds most of the herbivores and they in turn provide the food for the carnivores. Unfortunately again we have the problem of preservation (or rather non-preservation) for grass is not the best of potential fossils. I have no doubt, however, that the grass family appeared in the 'explosion' of the flowering plants at the beginning of Late Cretaceous times. I have been told (though I cannot find a reference) that the earliest members of the grass family were bamboos. Again we must consider whether this is not just a matter of preservation as I discussed in Chapter 2. Woody bamboos are obviously more likely to be fossilized than the soft, low green stuff which I used to cut every weekend (they won't let me now).

With the spread of grass over the land surfaces came the very rapid evolution of the horse family and other grazing animals. In due course the evolution and cultivation of cereals made possible the change in the habits of early people from hunting and gathering to settled agriculture and the rise of civilizations. One also cannot over-exaggerate the importance of a close grass cover on the land to the speed and severity of erosion. Grass quite suddenly changed one of the basic physical processes in geology. We cannot therefore accept a gradualistic approach to earth history for this reason if for no others.

I always wonder why palaeontologists pay so much attention to sudden extinctions and so little to sudden appearances, such as the spiny foram *Hantkenina* in the Eocene and (my own favourite) the distinctive brachiopod *Peregrinella* in the Early Cretaceous. These have no apparent ancestors but appear 'out of the blue' in many parts of the world with no 'mothers or fathers'. To me births are more interesting than deaths and this alone makes me incline towards the processes of the Earth rather than to those of the heavens.

We had a symposium at Durham in 1986 to discuss extinctions in the geological record, with contributions from specialists on every major group (Larwood 1988). Speaker after speaker could find no evidence to support the idea of asteroid impacts causing mass extinctions. Several made the joke that if it was caused by the arrival of something from outer space, 'How did my beasts know that it was coming?' In every group from spores to mammals, it was a matter of decline rather than sudden disaster. In my own contribution to that symposium (Ager 1988b) I emphasized the dangers of data bases. It is easy to produce peaks if you compartmentalize your data. Very often in studying brachiopods around the world, I am given no stratigraphical range more accurate than 'Cretaceous'. I am lucky if I get 'Upper Cretaceous'. I cannot help therefore but find an apparent mass extinction at the end of that period. This was well illustrated by Teichert (1986). He showed how the actual stratigraphical distribution of one ammonite family, the Hildoceratidae, was misinterpreted if plotted as a simple maximum range. The same point was made by Hoffman (1989).

I am always struck by the fact that even ammonites become very scarce as one approaches the top of the Cretaceous and of course there are no dinosaurs in the classic sections mentioned above. It is, in fact, very difficult to correlate the marine and the non-marine strata at this level. To be fair, however, I must admit that, as previously mentioned, I was almost convinced about the extinctions myself when we waded along the Waipara River, near Christchurch in New Zealand and Alexa Cameron showed me the exact point in the section where a plesiosaur (seen later in the museum) had been extracted from within a metre of the Cretaceous–Tertiary boundary (Figure 13.6).

Jaeger (1986) maintained that (to quote his English summary):

Iridium anomalies and extinctions of organisms in earth history are two fundamentally different phenomena. The alleged sudden mass mortality at the C/T boundary is an illusion. The diversity of the many animal groups that become extinct, declined stepwise before the end of the Cretaceous. The impact hypothesis is not necessary to account for the extinctions at the end of the period. The rise and decline of important fossil groups in the Cretaceous were closely linked to earth history, particularly the exceptional geological situation that was characteristic of the whole Upper Cretaceous. The Upper Cretaceous transgression with its resultant unusually equable warm climate and optimum conditions for organic evolution was succeeded by one of the largest regressions. This fundamental transformation of the face of the earth in turn caused widespread changes in the formerly favourable life conditions.

I find these views persuasive, except that all palaeontologists have their favourite period and favourite group and tend to be persuaded by the evidence they know best. Thus many of us would think of our own periods as the ones in which the most exciting things occurred!

I wander, however, from those alleged extra-terrestrial impacts. Along with the iridium came other features such as very widespread strained quartz grains and tiny glass globules which are thought to have been produced by impacts. Perhaps it was inevitable that similar anomalies should be found at other levels of 'mass extinctions'. Perhaps I am becoming a cynic in my old age, but I cannot help thinking that people find things that they expect to find. As Sir Edward Bailey (1953) said 'to find a thing you have to believe it to be possible'. Nevertheless, there does seem to be good evidence of such isotopic anomalies, including iridium and rare earths, in a core through the Permian–Triassic boundary in the Carnic Alps (Holser *et al.* 1989).

I am always a little sceptical about the major extinctions postulated at this level. Of the four main groups of Palaeozoic fossils, the graptolites had disappeared before the Permian, the trilobites were reduced to a mere handful, the conodonts sailed through and the brachiopods were not as much affected as is generally supposed. When I dealt (1965) with a major group of brachiopods for the monumental R. C. Moore *Treatise on Invertebrate Paleontology*, our distinguished editor inserted a paragraph about the major changes at the end of the Permian. I had to ask him to remove it again as I did not think it was true. Certainly the names changed and the classification changed; there were even changes in the names of the anatomical parts of the charming beasts. But so far as I could see the lineages went straight through. It is, I think, a matter of the psychology of palaeontologists rather than the tribulations of nature. Specialists who work on Upper Palaeozoic faunas almost never look at the Mesozoic and vice versa (Ager 1992b). Certainly some groups had a bad time at this moment in geological history, but others went through apparently unaffected. The changes at the end of the Permian were probably due to a decrease in salinity of the world's seas (as suggested by Fischer 1979) which may in turn be related to the vast salt deposits stored away in the evaporites of the Upper Permian, seen in so many parts of the world. Posenato has demonstrated (1990) the relative numbers of stenohaline and euryhaline forms across the Permo-Triassic boundary in the Italian Alps. It may also be significant that all round the world, from Argentina to Arizona, from Britain to Bulgaria and on to China, there was a sudden increase in clastic deposition in Early Triassic times, generally known in Europe as the *Buntsandstein* and coming from the new mountains formed at the end of Palaeozoic times (Ager 1981a, Chapter 1).

Recent evidence shows that the huge Manicouagan astrobleme in Quebec, some 65 km across, can be dated as 210 million years old, that is at about the Triassic–Jurassic boundary and may be blamed for the extinctions at that time, though I do not think that these were as general as is often supposed and I blame the obvious extinctions (such as the spiriferid brachiopods) on the widespread black shales that are found at that level. These were the work of Pluto deepening the depositional basins.

It must be said, however, that there were other mass extinctions which cannot be blamed on any particular external agency. Noe-Nygaard, Surlyk & Piasecki (1987) blamed mass mortality in an Early Cretaceous lagoon on Bornholm in the Baltic on dinoflagellate blooms and something similar is seen in a bed in the Oligocene of Romania, full of fish with their mouths open, gasping their last breaths. There also appears to have been a major extinction at the end of Ordovician times, but no evidence of iridium or other anomalies at the Ordovician–Silurian boundary. There is also the well-documented one at the end of the Frasnian Age in the Late Devonian and an anomaly has been found at this level in Australia. Nevertheless as Donovan pointed out (1987) it has not been found elsewhere although the extinctions at this level are a world-wide phenomenon.

I am particularly interested in the extinction of many large mammals only a few thousand years ago. This happened in Europe with such forms as the so-called 'Giant Irish Elk', but it is most clearly seen in the western United States, where six major species: the Columbian mammoth, the dire-wolf, the camel, the western bison, the giant armadillo and the horse (which was re-introduced by the Spaniards) all died out between 8000 and 7700 years ago (Newell 1963, 1967). Several more had disappeared in the previous few centuries (Figure 13.7). This was after the end of the Pleistocene glaciations, which do not appear to have caused extinctions except locally. The effects of the ice were largely to induce migrations. But the extinction of these large mammals was also, and probably more significantly, after the arrival of humans. Similarly, the giant ground sloth *Megatherium* only lasted in other parts of America long enough to be associated with early people. It has been suggested that waves of slaughter went with the arrival of humans on the different continents (Martin *et al*. 1967). I have further suggested that it was mankind armed with a new weapon, the Acheulian stone axe, that was to blame (Ager 1991a). It may be said that this was not a natural process, but the human was just one more species pursuing its selfish aims like all the other self-centred species in the history of the Earth. Certainly there were no extra-terrestrial 'bolides' to blame in this case. A later phase of slaughter by *Homo sapiens* came with the replacement of the bow and arrow by the rifle. Such was the near extinction of the American bison or 'buffalo'. It is difficult for us now to imagine the vast numbers of such beasts that formerly roamed the western prairies and could hold up a train for hours as they crossed the lines in front of it. It has been estimated that there were about 30 million in the American West when the first Europeans arrived there. With the work of 'Buffalo Bill' and his friends, this was reduced to a few hundred in the first half of this century and only a few now remain in small reserves such as that near Banff in the Canadian Rockies.

Several authors, notably Raup and Sepkoski (1984) claimed to recognize a cyclicity of 26 million years in extinctions based on the counting of numbers of taxa. This was usefully discussed by Jablonski (1985) and argued again by Raup & Sepkoski (1986). This also was not confirmed by the meeting of palaeontologists at Durham mentioned above. Again

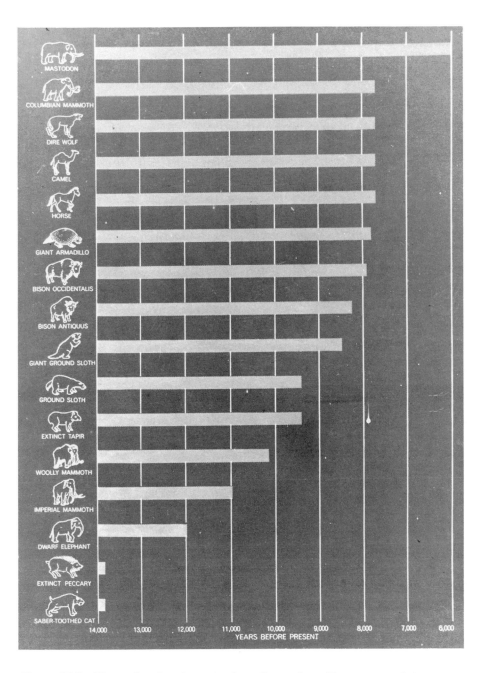

Figure 13.7   Figure showing the extinction of a number of large mammals in western North America, concentrated within a few hundred years. From Newell (1963) by kind permission of the author and *Scientific American*. The extinction of *Homo sapiens* would come slightly to the right of the diagram.

I must express my own scepticism, since families and genera are only fictions, invented by man for his own convenience and are highly subjective. They can be related to human geography and the distribution of taxonomic 'splitters'. Species alone have some sort of validity in the taxonomic hierarchy, since they represent interbreeding or potentially interbreeding individuals, which may be judged statistically; but even this is impossible to prove in fossil assemblages.

I should mention here other astronomical causes that have been put forward as possible causes for mass extinctions and other catastrophic events on Earth. One has been a 'wobble' in the Earth's rotation which may have caused the glaciations. There has been much talk of solar flares and cosmic dust. A recent discovery was the bursts of gamma rays all over the universe, though no-one I think has yet blamed terrestrial disasters on this phenomenon. Stothers (1988) suggested a correlation between impact craters on the Earth and Oort's Comet cloud.

There is also the suggestion of a dark partner to the Sun, appropriately called 'Nemesis', which periodically passed in front of our private star with periodic disastrous results for the inhabitants of the Earth, as in the cycles of extinctions just mentioned. I should not criticize, however, as I have played the numbers game myself in the Mesozoic (Ager 1981b), recognizing cycles at roughly 30 million year intervals. These were not in fossil extinctions, but in the repetition of transgressions, climatic optima and regressions, as far as possible on a world-wide basis. However, I did not attribute this to any happenings outside the Earth, but to variations in the rate of ocean floor spreading which was controlled by internal processes.

So we may now turn to Pluto and to forces within the Earth and on its surface. Earthquakes and hurricanes and many of the other processes I have discussed in this book can only be blamed on the Earth itself. I would say the same of volcanism, though it has been suggested that impacts caused the great outpourings of plateau basalts (Baksi 1990). On the other hand many of the phenomena discussed earlier may have had a terrestrial cause. Thus high iridium anomalies have been found in volcanic ejecta, as mentioned above, and even shocked quartz grains have been regarded as a product of volcanism, though Owen & Anders (1988) did produce further evidence for their non-volcanic origin.

Only fairly recently did geologists recognize mantle plumes rising as 'hot spots' in the crust and causing volcanicity. They probably moved with time and can be traced across continents as can a belt of Tertiary volcanicity in Europe, which extends from the Apuseni mountains in Romania, via Hungary and Czechoslovakia to the Eifel district of Germany (Ager 1980, Figure 9.10). It is not clear whether the hot spot moved under the crust or the European plate moved over a hot spot.

The dust emitted from volcanoes is another possible cause of the disasters above which have been blamed on extra-terrestrial impacts.

Satellite pictures have shown that the recent eruption of Pinatubo (1991–2) in the Philippines sent a belt of sulphurous haze all round the world's tropics covering some 40% of the Earth's surface. It also probably caused global cooling. It may have triggered off the warm current known as El Niño on the west coast of South America, though this works in the opposite direction by episodically warming that region. It has been known before to have been associated with volcanic eruptions. The hypothesis of 'darkness at noon' as blamed by Ken Hsü (1986) on cosmic impacts may equally well be blamed on the clouds of dust produced by a major volcanic eruption. It has been generally reported that Pinatubo was responsible, helped by human-produced aerosols, for a reduction in the ozone layer over Europe. This may be unfortunate for our species, with the increased ultraviolet radiation causing skin cancers, but it could be argued that it is good for life as a whole, since it is likely to increase genetic mutations and therefore the evolution of new species.

Plate tectonics now seems to rule everything and that obviously is essentially an episodic matter. The whole business of magnetic striping of the ocean floors means that the plate movements did not happen continuously but at intervals. This is where episodicity in my terminology is another word for catastrophism. What is more, where the plate movements have been studied in detail, as in the Red Sea, my friend Peter Styles has shown me that there were even longer gaps in the spreading story (Girdler & Styles 1974). It would seem that Pluto slacked at times in stoking his subterranean fires. The magnetic reversals of plate tectonics have also been related to extinctions, notably of Radiolaria, but this has been disputed. Loper *et al.* (1988) related reversals in magnetic fields to climate and mass extinctions. They suggested that cycles of activity in the mantle and core caused intermittent volcanic activity and that in turn affected the climate. I have suggested (Ager 1981a, Chapter 8) how subduction could control transgressions and regressions and with them diachronous sedimentation and the evolution of life.

Where the plates are really active, as in the Rift Valleys of East Africa, the movements are very obvious. In the Afar Triangle of north-east Africa, cracks can be seen in the ground where the two sides of the Great Rift are still moving apart. It is said that if you put a stick across one of these cracks in the morning, by the evening it will have fallen in. The same rifting continues up into the Middle East and it was probably movement along one of the faults that brought down the walls of Jericho in biblical times. Chris Walley showed the faulting continuing into Lebanon (1988).

The rifting may also have affected the fauna and flora. We all know about the distribution of the Permian plant *Glossopteris* in the scattered parts of Gondwana, including Antarctica. There was also the Late Palaeozoic fresh-water reptile *Mesosaurus* which is found both in Africa and

South America. We also have examples in recent times, such as the flightless birds of the southern continents: the rhea in South America, the ostrich in Africa, the emu in Australia and the kiwi (and the recently extinct moa) in New Zealand. The beautiful antelope the oryx, with its graceful curved back horns has an interesting distribution. The Arabian oryx, *Oryx gazella gazella* was separated by the splitting open of the Red Sea and Gulf of Aden from the closely related subspecies *Oryx gazella beisa* (the Beisa oryx) which lives in the 'Horn' of Africa. Similarly the reindeer of arctic Europe and Asia was separated from the caribou of arctic North America. They are usually regarded as the same species (*Rangifer tarandus*), but differ slightly in that the variety on the east side of the Atlantic is smaller and has shorter legs. They have also been domesticated or semi-domesticated since the fifth century, but that is man interfering again. There are many examples in the fossil record of the part played by Pluto and his convection currents in separating and bringing together faunas. In my own field Upper Jurassic 'Ethiopian' brachiopods such as *Septirhynchia* and *Bihenithyris* extend from east Africa to north-west Africa. Some are found in Israel and in north-west India but not in Asiatic Turkey or Europe; other brachiopods, such as the Early Jurassic *Spiriferina* and the Early Cretaceous *Peregrinella* managed to cross the Pacific (Ager 1986c, 1988b). All these things can be blamed on Pluto and his convection currents, as can the isolation of Australia and the evolution of its remarkable diversity of marsupials whilst the placental mammals prospered in the rest of the world.

The middle of the Atlantic continues to split open as Pluto stokes his fires and produces the continuing volcanic activity in Iceland, with the formation of the new volcanic island of Heimay. Southwards along the split, some thirty years ago between October 1961 and March 1962 (van Padang *et al.* 1967), the eruption on the South Atlantic island of Tristan da Cunha necessitated the temporary evacuation of the small population to England. Subsequently they decided that they would rather go back to their lonely island (in spite of the volcano)! So things continue to happen and when they do so at the present day, they are sudden and catastrophic in geological terms.

One just cannot have a gradual earthquake, a gradual hurricane, a gradual storm surge, a gradual magnetic reversal, a gradual volcanic explosion or – for that matter – a gradual impact. Though the process of subduction may seem slow and gradual, there can only be one moment when the continents collide. It now seems that we do not often see the gradual emergence of a new species either. All these things are sudden and, in a sense, catastrophic.

As I have said previously, I am very much inclined towards the views of Lovelock (1979, 1983, 1988, 1989) and his concept of the Earth as *Gaia*, like a huge working organism, with everything related to everything else, without the intrusion of outside bodies. In a way this was not a new idea; we have known almost since the beginning of science that the plants produced the oxygen needed by the animals and the animals produced the carbon dioxide needed by the plants. In a number of papers my good friend Jan Brunn of Paris has related the evolution of the atmosphere to volcanism, life and sedimentation (see for example Brunn 1983). In this way I see the periodic global warming in the past, with the increase of carbon dioxide in the atmosphere, as related to carbonate deposition in the seas and the spread of limestones across the continents as I discussed in Chapter 7.

In Chapter 8 of my book *The Nature of the Stratigraphical Record* (Ager 1981a) I suggested how subsidence, diachronous sedimentation and plate subduction could provide the control for the record as we see it. This again was a matter of internal, not external, forces. It is also the internal forces of the Earth that are most obvious to us as violent processes today, whether they be volcanic eruptions or earthquakes. At the north end of the great East African rift valley, in the Afar Triangle, cracks are perceptibly opening in the ground, as mentioned above. Farther north was the opening of the Red Sea and the continuation of that fracture into Lebanon, the effects of which (such as the subspeciation of the oryx and the fall of the walls of Jericho I have already discussed). The Arabian peninsula and Africa were probably joined recently enough to allow early humans to cross from Africa into Asia and thence to populate the rest of the world. If we go farther back in time we have the similar splitting of Madagascar from the African mainland and the isolation of its distinctive fauna and flora as discussed in Chapter 10.

In conclusion I must say that I do not reject the importance of the impact hypothesis. The case for impacts by extra-terrestrial bodies has been strengthened in my mind by the news that about 100 asteroids have been found crossing the Earth's orbit, mostly in the last five years, and that in January this year (1991) the asteroid 1991 BA came within 170 000 km of the Earth, that is about half the distance to the Moon. It has been calculated that a major impact can be expected once every 300 000 years (Verchnur 1991). Indeed the US Congress has recently recommended the setting up of a £40 million global ring of observatories to provide an early warning of the next extra-terrestrial intruder, in the hope that it may be deflected by firing nuclear missiles at it. This is not science fiction! But if it is a matter of choosing between heavenly processes and subterranean ones, I must declare myself on the side of the devils rather than that of the angels. I feel bound to quote the song from '*Cymbeline*': 'Fear no more the lightning flash, Nor the all-dreaded thunder-stone'.

Nevertheless, it is obvious to me that the whole history of the Earth is one of short, sudden happenings with nothing much in particular in between. I have often been quoted for my comparison of Earth history with the traditional life of a soldier, that is 'long periods of boredom separated by short periods of terror'. Perhaps a more everyday comparison would be with the life of a goalkeeper in soccer. For the benefit of my American friends I should explain that he is the lonely man waiting to defend the goal when the other side gets near. Unlike those other strange games such as rugby 'football' and American 'football', in which paradoxically, the foot is only occasionally used, the goalkeeper in soccer is the only man who is allowed to use his hands on the field. I was a goalkeeper before I was a soldier and well remember the long periods of chilly boredom interspersed with sudden threats of catastrophe. That is how I see the history of our planet. How then will it end?

The British philosopher Arthur Balfour, who became Prime Minister at the beginning of this century, wrote: 'The energies of our system will decay, the glory of the Sun will be dimmed, and the Earth, tideless and inert, will no longer tolerate the race which has for a moment disturbed its solitude. Man will go down into the pit, and all his thoughts will perish'. A similarly pessimistic note is struck in T. S. Eliot's famous concluding lines in 'The Hollow Men':

This is the way the world ends
This is the way the world ends
This is the way the world ends
Not with a bang but a whimper.

If we are thinking of 'the world' as the solid globe on which we live, then in spite of that great poet, it will probably end with a bang as it falls back into the Sun or is shattered in a collision with another astronomical body. But if we are thinking of 'the world' in our usual egocentric way as meaning *Homo sapiens* and our personal interests, then unless we destroy ourselves by our own folly, and long before that physical apocalypse our descendents will see a gradual decline in the organic world, hurried on by a series of catastrophes, many of them of our own making. As in the past, the first to go will be the large animals such as the whales and elephants, which are already very vulnerable. As we continue to pollute the land, sea and atmosphere, more and more species will disappear of varying importance (economic or aesthetic) to ourselves. The land will become more and more over-populated by us and our domestic animals. The sea will be over-fished and the whole complex ecosystem will whimper to its end. We shall certainly be outlived by the bacteria and blue-green algae, but that is only fair as they were here first.

Though the physical world will end with a bang, the organic world will probably end with a whimper.

References

Agassiz, E. C. (edit.) 1886. *Louis Agassiz his life and correspondence*. Houghton, Mifflin & Co., Boston & New York, 2 vols, 794 pp.

Ager, D. V. 1955a. Field Meeting in the Central Cotswolds. *Proc. Geol. Ass.*, **66** (4), 356–65.

Ager, D. V. 1955b. Summer Field Meeting in Italy. *Proc. Geol. Ass.*, **66** (4), 329–52.

Ager, D. V. 1956. Geographical factors in the definition of fossil species. *In* P. C. Sylvester-Bradley (edit.) *The species concept in palaeontology*, Systematics Assoc., Publ. No. **2**, 105–9.

Ager, D. V. 1958. On some Turkish sediments. *Geol. Mag.*, **95**, 83–4.

Ager, D. V. 1963. *Principles of Paleoecology*. McGraw-Hill, New York, 371 pp.

Ager, D. V. 1965a. The adaptation of Mesozoic brachiopods to different environments. *Palaeogeog., Palaeoclimat., Palaeoecol.*, **1**, 143–72.

Ager, D. V. 1965b. Mesozoic and Cenozoic Rhynchonellacea. *In* Moore, R. C. (edit.) *Treatise on Invertebrate Paleontology*, Part H, Brachiopoda, **2**, H597-H625.

Ager, D. V. 1967. Brachiopod Palaeoecology, *Earth-Sci. Rev.*, **3**, 157–79.

Ager, D. V. 1969. Alleged sexual dimorphism in Mesozoic brachiopods. *In* Westermann, G. E. G. (edit.) *Sexual dimorphism in Fossil Metazoa and Taxonomic Implications*, Internat. Union geol. Sci., Prague, Ser. a, No. **1**, 34–6.

Ager, D. V. 1970. The Triassic System in Britain and its stratigraphical nomenclature. *Quart. Jl. geol. Soc., Lond.*, **126**, 3–17.

Ager, D. V. 1972. Summer Field Meeting in Bulgaria. *Proc. Geol. Ass.*, **83** (3), 239–68.

Ager, D. V. 1973. *The Nature of the Stratigraphical Record* (1st edn). Macmillan/Wiley, 114 pp.

Ager, D. V. 1974. Storm Deposits in the Jurassic of the Moroccan High Atlas. *Palaeogeog., Palaeoclimat., Palaeoecol.*, **15**, 83–93.

Ager, D. V. 1975a. The Jurassic World Ocean (with special reference to the North Atlantic). Keynote Address, Jurassic Northern North Sea Symposium, Norwegian Petroleum Exploration Soc., 1–1 – 43.

Ager, D. V. 1975b. The Geological Evolution of Europe. Presid'l Address, *Proc. Geol. Ass.*, **86** (2), 127–54.

Ager, D. V. 1976. The nature of the fossil record. Presid'l Address, *Proc. Geol. Ass.*, **87**, 131–60.

Ager, D. V. 1980. *The Geology of Europe*. McGraw-Hill Book Co., Maidenhead, 535 pp.

Ager, D. V. 1981a. *The Nature of the Stratigraphical Record* (2nd edn). Macmillan/Wiley, Basingstoke, 122 pp.

Ager, D. V. 1981b. Major marine cycles in the Mesozoic. *Jl. geol. Soc., Lond.*, **138** (2), 159–66.

Ager, D. V. 1983. Allopatric speciation – an example from the Mesozoic Brachiopoda. *Palaeontology*, **26** (3), 555–65.

Ager, D. V. 1985a. Beaches that I never stormed. *Tank (Jl. roy. Tank Regt.)*, **67**, No. 692, 93–6.

Ager, D. V. 1985b. Pollution? It's only natural. *New Sci.*, No. 1542, 80.

Ager, D. V. 1985c. The ore that launched a thousand ships. *New Sci.*, No. 1461, 28–9.

Ager, D. V. 1986a. Evolutionary patterns in the Mesozoic Brachiopoda. *In* Racheboeuf, P. R. & G. C. C. Emig (edits). *Les Brachiopodes fossiles et actuels*. Proc. 1st internat. Congr. Brachiopoda, Biostratigraphie du Paléozoique, Brest, **4**, 33–41.

Ager, D. V. 1986b. A reinterpretation of the basal 'Littoral Lias' of the Vale of Glamorgan. *Proc. Geol. Ass.*, **97** (1), 29–35.

Ager, D. V. 1986c. Migrating Fossils, Moving Plates and an Expanding Earth. Presd'l Address, Sect. C, Brit. Assoc. Adv. Sci., *Modern Geol.*, **10**, 377–90.

Ager, D. V. 1988a. Cosmic collisions and Earth history – a geolologist's view. *Jl. brit. astron. Assoc.*, **98** (1), 85–8.

Ager, D. V. 1988b. Extinctions and survivals in the Brachiopoda and the dangers of data bases. *In* Larwood, G. P. (edit.), *Extinction and Survival in the Fossil Record*, Systematics Assoc./Clarendon Press, 365 pp.

Ager, D. V. 1989. Lyell's Pillars and Uniformitarianism. *Jl. geol. Soc. Lond.*, **146**, 603–5.

Ager, D. V. 1991a. Ban the stone axe! *New Sci.*, No. 1752, 61–2.

Ager, D. V. 1991b. It's a pity about the panda. *New Sci.*, No. 1771, 56–7.

Ager, D. V. 1992a. The myth of the cave man. *New Sci.*, No. 1806, 56.

Ager, D. V. 1992b. Minor extinction. *New Sci.*, No. 1809, 53.

Ager, D. V. 1993. The Nature of the Stratigraphical Record (3rd edn). Wiley, Chichester. *In the press.*

Ager, D. V., A. Childs & D. A. B. Pearson 1972. The evolution of the Mesozoic Rhynchonellida. *Geobios*, **5** (2–3), 157–235.

Ager, D. V., D. T. Donovan, W. J. Kennedy, W. S. McKerrow, D. C. Mudge & B. W. Sellwood 1973. The Cotswold Hills. *Geol. Ass. Guide* No. 36, 34 pp.

Ager, D. V. & D. Edwards 1986. The Fauna and Flora of the Rhaetian of South Wales and Adjacent Areas. *Nature in Wales*, **4**, 71–9.

Ager, D. V. & B. D. Evamy 1963. The Geology of the Southern French Jura. *Proc. Geol. Ass.*, **74** (3), 325–56.

Ager, D. V. & P. Wallace 1966a. The environmental history of the Boulonnais, France. *Proc. Geol. Ass.*, 77, 385–417.

Ager, D. V. & P. Wallace 1966b. Flume experiments to test the hydrodynamic properties of certain spiriferid brachiopods with reference to their supposed life orientation and mode of feeding. *Proc. geol. Soc., Lond.*, No. 1635, 160–3.

Ager, D. V. & P. Wallace 1971. The distribution and significance of trace fossils in the uppermost Jurassic rocks of the Boulonnais, northern France. *In* T. P. Crimes & J. C. Harper (edits), *Trace Fossils, Geol. Jl.*, Spec. Issue No. 3, 1–18.

Albritton, C. C. 1989. *Catastrophic episodes in earth history*. Chapman & Hall, London, 221 pp.

Almeras, Y. & A. Boullier 1990. Evolutionary patterns in the terebratulid genus *Caryona* Cooper (Brachiopoda, Jurassic of France). *Historical Biol.*, **4**, 15–37.

Alvarez, L. W., W. Alvarez, F. Asaro & H. V. Michell 1979. Anomalous iridium levels at the Cretaceous/Tertiary boundary at Gubbio, Italy: negative results of a test for a supernova origin. *In* W. K. Christensen & T. Birkelund (edits), *Cretaceous-Tertiary Boundary Events*, University Copenhagen Symp. vol. 2, 50–3.

Alvarez, L. W., W. Alvarez, F. Asaro & H. V. Michell 1980. Extraterrestrial cause for the Cretaceous-Tertiary extinction. *Science*, **208**, 1095–108.

Alvarez, L. W., W. Alvarez, F. Asaro & H. V. Michell 1984. The end of the Cretaceous. Sharp boundary or gradual transition? *Science*, **223**, 1183–6.

Ambraseys, N. N. 1962. Data for the investigation of the seismic sea-waves in the eastern Mediterranean. *Bull. seismol. Soc. Amer.*, **52** (4), 895–913.

Ambraseys, N. N. 1968. Early earthquakes in north-central Iran. *Bull. seismol. Soc. Amer.*, **58** (2), 485–96.

Ambraseys, N. N. 1970. A note on an early earthquake in Macedonia. Proc. 3rd. europ. Sympos. Earthquake Engineering, *Bulgar. Acad. Sci.*, 73–78.

Ambraseys, N. N. 1971. Value of historical records of earthquakes. *Nature*, **232**, 375–8.

Anderson, H. 1987. Is the Adriatic an African promontory? *Geology*, **15**, 212–5.

Anderson, I. 1990. Darwin may founder on the Great Barrier Reef. *New Sci.*, No. 1739, 15.

Andrews, M. 1982. *The flight of the Condor*. Collins & Brit. Broadcasting Corp., 158 pp.

Anon. 1981. Basaltic volcanism, Italy Project. *In Basaltic Volcanism on the Terrestrial Planets*. Pergamon Press, 974–87.

Anon. 1982. The Channeled Scablands of Eastern Washington. US Dept of the Interior, Geological Survey, US Govt Printing Office, 25 pp.

Anon. 1991. Glass beads support dinosaur theory. *New Sci.*, No. 1774, 23.

Bagge-Johansen, M. 1988. Brachiopod extinctions in the Upper Cretaceous to lowermost Tertiary Chalk of northwest Europe. *Rev. españ. Paleont., No. extraord.*, Palaeontology and Evolution: Extinction Events, 41–56.

Bagge-Johansen, M. 1989a. Adaptive radiation, survival and extinction of brachiopods in the northwest European Upper Cretaceous-Lower Paleocene Chalk. *Palaeogeog., Palaeoclimat., Palaeoecol.*, **74**, 147–204.

Bagge-Johansen, M. 1989b. Background extinction and mass extinction of the brachiopods from the Chalk of northwest Europe. *Res. Rept., Soc. econ. Paleontologists Mineralogists*, 243–50.

Bailey, E. B. 1953. Some features of Provençal tectonics. *Quart. Jl. geol. Soc., Lond.*, **108** (2), 135–55.

Bailey, E. B. & J. Weir 1932. Submarine faulting in Kimmeridgian times: east Sutherland. *Trans. roy. Soc. Edinb.*, **57**, 429–67.

Baksi, 1990. Timing and duration of Mesozoic-Tertiary flood-basalt volcanism. *Eos*, **71** (49), 1835–40.

Barnes, J. 1989. *A History of the World in 10½ Chapters*. Picador/Pan Books/ Jonathan Cape, 309 pp.

Bassett, M. G. & D. E. Edwards 1982. Fossil plants from Wales. *National Mus. Wales geol. ser.*, Cardiff, 42 pp.

Berggren, W. A. & J. A. Van Couvering (edits). 1984. *Catastrophes and Earth History*. Princeton Univ. Press, 463 pp.

Berner, R. A. 1991. A model for atmospheric CO_2 over Phanerozoic time. *Amer. Jl. Sci.*, **291**, 339–76.

Bini, A., M. B. Cita & M. Gaetani 1978. Southern alpine lakes. Hypothesis of an erosional origin in the Messinian entrenchment. *Marine Geol.*, **27**, 271–88.

Boucot, A. 1953. Life and death assemblages among fossils. *Amer. Jl. Sci.*, **251**, 25–40 (errata p. 248).

Bradley, W. H. 1948. Limnology and the Eocene lakes of the Rocky Mountain region. *Bull. geol. Soc. Amer.*, **59**, 635–48.

Bretz, J. H. 1923. The channeled scablands of the Columbia Plateau. *Jl. Geol.*, **31**, 617–49.

Bretz, J. H. 1959. Washington's Channeled Scabland. *Washington Division of Mines and Geology. Bull.*, **45**, 57 pp.

Bretz, J. H. 1969. The Lake Missoula floods and the channeled scabland. *Jl. Geol.*, 77, 505–43.

Bridges, M. 1987. Classic landforms of the Gower Coast. *Geog. Assoc. Classic Landform Guides*, No. 7, 48 pp.

Broadhurst, F. M. & D. H. Loring 1970. Rates of sedimentation in the Upper Carboniferous of Britain. *Lethaia*, **3**, 1–9.

Broadhurst, F. M., I. N. Simpson & P. G. Hardy 1980. Seasonal sedimentation in the Upper Carboniferous of England. *Jl. Geol.*, **88**, 639–51.

Brooks, R. R., R. D. R. D. Reeves, R. D., X-H. Yang, D. E. Ryan, J. Holzbecher, J. D. Collen, V. E. Neall & J. Lee 1984. Element Anomalies at the Cretaceous-Tertiary Boundary, Woodside Creek, New Zealand. *Amer. Assoc. Adv. Sci.*, **226**, 539–42.

Brunn, J. H. 1983. Essai sur l'évolution de l'atmosphère; ses rapports avec la volcanisme, la vie, la sédimentation. *Bull. Soc. géol. France*, **25** (1), 117–28.

Buffetaut, E. 1987. *A short history of vertebrate palaeontology.* Croom Helm, London etc., 223 pp.

Bulow, K. von 1960. Planetarische Fundamentaltektonik dargelegt an Mond und Mars. *Rept. internat. geol. Congr.,* Norden, Copenhagen, Pt. **21**, 7–14.

Cadee, G. C. 1976. Sediment reworking by *Arenicola marina* on tidal flats in the Dutch Wadden Sea. Netherlands. *Jl. Sea Research,* **10**, 440–60.

Carey, S. W. 1988. *Theories of the Earth and Universe. A History of Dogma in the Earth Sciences.* Stanford Univ. Press, 413 pp.

Casey, R. 1968. The type-section of the Volgian Stage (Upper Jurassic) at Gotodische, near Ulyanovsk, USSR. *Proc. geol. Soc. Lond.,* No 1648, 14–5.

Chaloner W. G. & G. T. Creber 1990. Do fossil plants give a climatic signal? *Jl. geol. Soc. Lond,* **147**, 343–50.

Chamberlain, C. J. 1965. *The living cycads.* Hafner, New York & London, 171 pp. (facsimile of the 1919 edition).

Chevrier, R. M. 1990. Lake Nyos: phenomenology of the explosion event of December 30, 1986. *Jl. volcanol. & geothermal Res.,* **42** (4), 387–90.

Cita, M. B. 1982. The Messinian salinity crisis in the Mediterranean: a review. *In* Berkmer, H. & K. Hsu (edits), *Alpine-Mediterranean Geodynamics,* Geodynamics Ser., 7, 113–40.

Cita, M. B., A. Bini & C. Corselli 1990. Superfici erosione messiniane: una ipotsi sull' origine del laghi sud-alpini. *In* Barbanti, L., G. Giussani & R. De Bernardi (edits.), Lago Maggiore dalla ricerca alla gestione *Doc. Istitut. ital. Idrobiol.,* Palianza, **22**, 33–54.

Clauzon, G. 1979. Le canyon messinien de la Durance (Provence, France): une preuve paléogéographique du bassin profond de dessiccation. *Palaeogeog., Palaeoclimat., Palaeoecol.,* **29** (1/2), 15–40.

Clayton, R. N. & G. R. Stevens 1968. Palaeotemperatures of the New Zealand Jurassic and Cretaceous. *Tuatara,* **16** (1), 3–7.

Clube, S. V. M. 1978. Does our galaxy have a violent history? *Vistas in Astronomy,* **22**, 77.

Clube, S. V. M. & W. M. Napier 1982. *The Cosmic Serpent. A catastrophist view of Earth History.* Faber & Faber, 299 pp.

Copper, P. 1986. Frasnian/Famennian mass extinction and cold-water oceans. *Geology,* **14**, 835–39.

Crawford, A. R. 1979. The myth of a vast oceanic Tethys, the India-Asia problem and Earth expansion. *Jl. Petrol. Geol.,* **2**, 3–9.

Creber, G. T. & W. G. Chaloner 1985. Tree growth in the Mesozoic and Early Tertiary and the reconstruction of palaeoclimates. *Palaeogeog., Palaeoclimat., Palaeoecol.,* **52**, 35–60.

Crichton, M. 1991. *Jurassic Park.* Arrow Books, London, 401 pp.

Crocket, J. H., Officer, C. B., Wezel, F. C. & G. D. Johnson 1988. Distribution of noble metals across the Cretaceous/Tertiary boundary at Gubbio, Italy: iridium variation as a constraint on the duration and nature of Cretaceous/Tertiary boundary events. *Geology,* **16**, 77–80.

Cullen, D. J. 1980. Distribution, composition and age of submarine phosphorites on Chatham Rise, east of New Zealand. Spec. Publ. No. 29, *Soc. econ. Paleont. Mineralog.*, 139–48.

Cuvier, G. 1812–1820. *Recherches sur les ossemens fossiles de quadrupèdes, ou l'on retablit les caractères de plusieurs espèces d'animaux que les révolutions du globe paroissent avoir détruits.* Paris. (in 5 vols, the last in 2 parts).

Cuvier, G. 1827. *'Essay on the Theory of the Earth'.* (5th English edn). Blackwell & Cadell, Edinburgh & London, 550 pp.

Cuvier, G. & A. Brongniart 1822. *Description géologique des environs de Paris.* (new edn.) Dufour & d'Ocagne, Paris, 428 pp.

Darwin, C. 1851–5. A Monograph of the Fossil Cirripedes of Great Britain (Lepadidae, Balanidae, Verrucidae) *Monogr. Palaeontogr. Soc.* pt. I, i–iv, p. i–v; pt II, 1–44, pl. i–ii

Darwin, C. 1859. *On the Origin of Species by Means of Natural Selection.* (1st edn) John Murray, London, 490 pp.

Darwin, C. 1874. *The structure and distribution of coral reefs.* (2nd edn). Smith, Elder & Co., London, 278 pp.

Darwin, C. 1881. *The Formation of Vegetable Mould, through the Action of Worms.* John Murray, London, 326 pp.

Davies, K. 1991. The mystery of motor neuron disease. *New Sci.*, No. 1782, 21–5.

De Laubenfels, M. W. 1956. Dinosaur extinction: one more hypothesis. *Jl. Paleont.*, **30**, 207–18.

Dolan, R. & H. Lins, 1985. The Outer Banks of North Carolina. *US geol. Surv., Prof. Paper* 1177–13, 101 pp.

Donovan, S. K. 1987. Iridium anomalous no longer? *Nature*, **326**, 331–2.

Doust, H. 1968. *Palaeoenvironment studies in the Miocene Libya/Australia*, vol. 1. Unpubl. thesis, Univ. London (Imperial College), 254 pp.

Drake, C. L. & Y. Herman 1988. Did the dinosaurs die out or evolve into red herrings? *Northwest Sci. USA*, **62** (3), 131–46.

Drost-Hansen, W. & H. W. Neill 1955. Temperature anomalies in the properties of liquid water. *Phys. Rev.*, **100**, 1800.

Duncan, R. A. & D. G. Pyle 1988. Rapid extrusion of the Deccan flood basalts at the Cretaceous/Tertiary boundary. *Nature*, **333**, 841–3.

Edwards, D., K. L. Davies & L. Axe 1992. A vascular conducting strand in the early land plant *Cooksonia*. *Nature*, **357**, 683–5.

Ekman, S. 1953. *Zoogeography of the sea.* Sidgwick & Jackson, 417 pp.

Einsele, G. & A. Seilacher 1982. *Cyclic and Event Stratification.* Springer-Verlag, 536 pp.

Eldredge, N. 1971. The allopatric model and phylogeny in Paleozoic invertebrates. *Evolution*, **25**, 156–67.

Eldredge, N. & S. J. Gould 1972. Punctuated equilibria: an alternative to phyletic gradualism. *In* Schopf, T. J. M. (edit.), *Models in Paleobiology*, Freeman & Cooper, San Francisco, 82–115.

Ericson, D. B., G. Wollin & J. Wollin 1954. Coiling direction of *Globorotalia truncatulinoides* in deep-sea cores. *Deep Sea Research*, **2**, 152–8

Evans, G. 1965. Intertidal flat sediments and the environments of deposition in the Wash. *Quart. Jl. geol. Soc., Lond.,* **121**, 209–45.

Ferguson, J. 1978. Some aspects of the ecology and growth of the Carboniferous gigantoproductids, *Proc. Yorks. geol. Soc.,* **42**, 41–54.

Finckh, P. 1978. Are the southern alpine lakes former messinean canyons? – Geophysical evidence for preglacial erosion in the southern alpine lakes. *Marine Geol.,* **27**, 289–302.

Fischer, A. G. 1979. Rhythmic changes in the outer earth. *Newsl. geol. Soc. Lond.,* **8** (6), 2–3.

Fletcher, D. 1984. *Vanguard of Victory.* HMSO, London, 88 pp.

Fox, D. L., S. C. Crane & B. H. McConnaughey 1948. A biochemical study of the marine annelid worm *Thoracophelia mucronata Jl. mar. Res.,* 7, 567–686.

Franzen, J. L. 1985. Exceptional preservation of Eocene vertebrates in the lake deposit at Grube Messel (West Germany). *Phil. Trans. roy. Soc., Lond.,* **B311**, 181–6.

Franzen, J. L., J. Weber & M. Wuttke 1982. Senckenberg-Grabungen in der Grube Messel bei Darmstadt, 3 Ergebnisse 1979–1981. *Courier Forschungsinstitut Senckenberg,* **54**, 1–118.

Freeth, S. J. 1990. The anecdotal evidence, did it help or hinder investigation of the Lake Nyos gas disaster? *Jl. Volcanol. & Geotherm. Res.,* **42**, 373–80.

Freeth, S. J. & R. L. F. Kay 1987. The Lake Nyos gas disaster. *Nature,* **325**, 104–5.

Freeth, S. J. *et al.* 1990. Conclusions from the Lake Nyos disaster. *Nature,* **348**, 201.

Froggatt, P. C., C. S. Nelson, L. Carter, G. Griggs & K. P. Black 1986. An exceptionally large Late Quaternary eruption from New Zealand. *Nature,* **319**, 578–82.

Furness, R. W. 1988. Predation on ground-nesting seabirds by island populations of red deer *Cervus elephas* and sheep *Ovis. Jl. Zool.,* **216**, 565–73.

Gaillard, C., J-P. Bourseau, M. Boudeulle, P. Pailleret, M. Rio & M. Roux 1985. Les pseudo-biohermes de Beauvoisin (Drôme): un site hydrothermal sur la marge téthysienne à l'Oxfordien? *Bull. Soc. geol. France,* Ser. **8**, 1, 69–78.

Gaillard, C. & Y. Rolin 1986. Paléobiocoenoses susceptibles d'être liées à des sources sous-marines en milieu sédimentaire. L'example des pseudobiohermes des Terres Noires (SE France) et des tepée buttes de la Pierre Shale Formation (Colorado, USA). *C. R. Acad. Sci. Paris,* **303**, Ser. II, 1503–8.

Gaillard, C. & Y. Rolin 1988. Relation entre tectonique synsédimentaire et pseudobiohermes (Oxfordien de Beauvoisin-Drôme-France). Un argument supplémentaire pour interpréter les pseudobiohermes comme formé au droit de sources sous-marines. *C. R. Acad. Sci. Paris,* **307**, Ser. II, 1265–70.

Gasse, F. & J. C. Fontes 1989. Palaeoenvironments and palaeohydrology of a typical closed lake (Lake Asai, Djibouti) since 10 000 yr. BP. *Palaeogeog., Palaeoclimat., Palaeoecol.,* **69**, 67–102.

Girdler, R. W. & P. Styles 1974. Two stage Red Sea floor spreading. *Nature*, **247**, 7–11.

Glanvill, W. S. 1981. *Quantitative palaeoecological studies with special reference to Early Carboniferous rocks, Mid Glamorgan*. Unpubl. Ph.D. thesis, Univ. Wales (Swansea), 372 pp.

Goldring, R. 1962. The trace fossils of the Baggy Beds (Upper Devonian) of North Devon, England. *Palaeont. Zeitschrift*, **36**, 232–52.

Gould, S. J. 1966. Allometry and size in ontogeny and phylogeny. *Biol. Rev., Cambridge philos. Soc.* **41**, 587–640.

Gould, S. J. 1972. Allometric fallacies and the evolution of *Gryphaea*. A new interpretation based on White's criterion of geometric similarity. *Evolutionary Biology*, **6**, 91–118.

Gould, S. J. 1974. The origin and function of 'bizarre' structures: antler size and skull size in the 'Irish Elk', *Megaloceros giganteus. Evolution*, **28** (2), 191–220.

Gould, S. J. 1980. *The Panda's Thumb*. W. W. Norton & Co. (Paperback edn Penguin Books, 1985), 285 pp.

Gould, S. J. 1987. *Time's Arrow Time's Cycle*. Harvard Univ. Press, Cambridge Mass. & London, 222 pp.

Gould, S. J. & Eldredge, N. 1977. Punctuated Equilibria: the tempo and mode of evolution reconsidered. *Paleobiol.*, **3**, 115–55.

Graedel, T. E., I-J. Sackmann & A. I. Boothroyd 1991. Early solar mass loss: a potential solution to the weak sun paradox. *Geophys. Res. Letters*, **18** (10), 1881–4.

Gregory, J. 1988. Unsung champion of science. *The Independent* (newspaper), 7 March 1988.

Gretener, P. E. 1967. Significance of the rare event in geology. *Bull. amer. Assoc. Petrol. Geol.*, **51**, 2197–206.

Hallam, A. 1959. On the supposed evolution of *Gryphaea* in the Lias. *Geol. Mag.*, **96**, 99–108.

Hallam, A. 1984. Asteroids and extinctions – no cause for concern. *New Sci.*, No. 1428, 30–2.

Hallam, A. 1989. The case for sea-level change as a dominant causal factor in mass extinction of marine invertebrates. *Phil. Trans. roy. Soc. Lond.*, B, **325**, 437–55.

Hamblin, W. K. 1990. Late Cenozoic lava dams in the western Grand Canyon. *In* Beus, S. S. & N. Morales (edits), *Grand Canyon Geology*, Oxford Univ. Press/Museum North Arizona Press, 385–433.

Hamilton, Sir W. 1774. *Observations on Mount Vesuvius, Mount Etna and other volcanos in a series of letters addressed to the Royal Society*. T. Cadell, London. New Edition, 179 pp.

Hancock, J. M. 1975. The petrology of the Chalk. Presid'l Address, *Proc. Geol. Ass.*, **86**, 499–536.

Harrington. C. 1962. Search for the City of Raleigh. US Dept. Inter. National Park Service, 63 pp.

Hattin, D. E. 1986. Carbonate Substrates of the Late Cretaceous Sea, Central Great Plains and Southern Rocky Mountains, *Palaios*, V (1), 347–67.

Hayes, M. O. 1967. Hurricanes as geological agents: case studies of Hurricane Carla, 1961 and Cindy, 1963. *Bur. econ. Geol. Univ. Texas, Rept.* Investigation No. 61, 56 pp.

Hecht, J. 1990. Did a comet end the Triassic period? *New Sci.*, No. 1771, 25.

Herman, Y. 1988. Selective extinction of marine plankton at the end of the Mesozoic Era: the fossil and stable isotope record. *Abstr. Conf. Global catastrophes in Earth history*, Snowbird, Utah, 75.

Hickey, L. J. 1984. Changes in the angiosperm flora across the Cretaceous-Tertiary boundary. *In* Berggren & Van Couvering (q.v.), 279–313.

Hoffman, A. 1989. Mass extinctions: the view of a sceptic. *Jl. geol. Soc. Lond.*, **146**, 21–35.

Holland, J. D., 1984. *The chemical evolution of the atmosphere and oceans*. Princeton Univ. Press.

Holser, W. T. & 14 others 1989. A unique geochemical record at the Permian/Triassic boundary. *Nature*, **337**, 39–44.

Hooykas, R. 1984. Closing remarks. *In* Dudich, E. (edit.), *Contributions to the History of Geological Mapping*. INHIGEO Sympos., Akad. Kiado, Budapest, 423–29.

Hooper, P. R. 1982. The Columbia River Basalts. *Science* **215**, 1463–8.

Hough, J. L. 1958. *Geology of the Great Lakes*. Univ. Illinois Press.

Hsu, K. J. 1983. *The Mediterranean was a Desert*. Princeton Univ. Press, 197 pp.

Hsu, K. J. 1986. *The Great Dying*. Harcourt Brace Jovanovich, San Diego, 292 pp.

Hsu, K. J. 1990. Actualistic catastrophism and global change. *Palaeogeog., Palaeoclimat., Palaeoecol.* (Global & Planetary Change Sect.), **89**, 309–13.

Hsu, K. J. 1991. Ur-canyons im Suedtessin. *Heureka, Untersuch. Nationalforschungsprogrammes*, Nr. 20, 4 pp.

Hunt, A. P. 1991. Integrated vertebrate and plant taphonomy of the Fossil Forest area (Fruitland and Kirtland formations: Late Cretaceous), San Juan County, New Mexico, USA. *Palaeogeog., Palaeoclimat., Palaeoecol.*, **88**, 85–107.

Ivanova, E. A., T. M. Bleskaya & I. I. Chndinova 1962. Ecology and development of Silurian and Devonian brachiopods of the Kuznetsk, Minusinsk and Tuvinsk Basins. *Tr. Paleont. Inst. Akad. Nank. SSSR*, **102**, 1–226 (in Russian).

Jablonski, D. 1985. Marine regressions and mass extinctions: a test using the modern biota. *In* J. W. Valentine (edit.). *Phanerozoic diversity patterns*, Princeton Univ. Press, 335–54.

Jaeger, H. 1986. Die Faunenwende Mesozoikum/Känozoikum – nüchtern betrachtet. *Zeitschr. geol. Wis., Berlin*, **14** (6), 629–56.

Jarrett, R. D. & H. E. Malde 1987. Paleodischarge of the late Pleistocene Bonneville Flood, Sake River, Idaho, computed from new evidence. *Bull. geol. Soc. Amer.*, **99**, 127–34.

Jarzembowski, E. A. 1989. The occurrence and diversity of Coal Measure insects. *Jl. geol. Soc. Lond.*, **144**, 507–11.

Jeremiah 1611. The book of the prophet Jeremiah. 5, 22, *In* 'The Bible', Authorized Version (King James's), Cambridge Univ. Press, 708 or Oxford Univ. Press, 623 (in my undated editions).

Johnson, A. L. A. & C. D. Lennon 1990. Evolution of Gryphaeate oysters in the Mid-Jurassic of western Europe. *Palaeontology*, **33**, 453–85.

Johnson, J. G. 1975. Allopatric speciation in fossil brachiopods. *Jl. Paleont.*, **49**, 646–61.

Johnson, J. G. 1982. Occurrence of phyletic gradualism and punctuated equilibria through Paleozoic time. *Jl. Paleont.*, **56**, 1329–31.

Johnson, M. E. 1988. Why are ancient rocky shores so uncommon? *Jl. Geol.*, **96**, 469–80.

Johnson, M. E., D. F. Skinner & K. C. Macleod 1988. Ecological zonation during the carbonate transgression of a Late Ordovician rocky shore northeastern Manitoba, Hudson Bay, Canada. *Palaeogeog., Palaeoclimat., Palaeoecol.*, **65** (1/2), 93–114.

Jones, O. T. & W. J. Pugh 1949. An early Ordovician shore-line in Radnorshire, near Builth Wells. *Quart. Jl. geol. Soc. Lond.*, **105**, 65–99.

Kazmer, M. 1990. Birth, life and death of the Pannonian Lake. *Palaeogeog., Palaeoclimat., Palaeoecol.*, **79** (1/2) 171–88.

Kelling, G. 1968. Patterns of sedimentation in the Rhondda Beds of S. Wales. *Bull. amer. Ass. Petrol. Geol.*, **52**, 2369–86.

Kellogg, D. E. 1975. The role of phyletic change in the evolution of *Pseudocubus vema* (Radiolaria). *Paleobiol.*, **1**, 359–70.

Kerr, R. A. 1991. Did a volcano help to kill off the dinosaurs? *Science*, **252**, 1496–7.

King, A. F. 1965. Xiphosurid trails from the Upper Carboniferous of Bude, N. Cornwall. *Proc. geol. Soc. Lond.*, No. 1626, 162–5.

Kirby, R. 1987. Sediment exchanges across the coastal margins of NW Europe (introduction to Coastal Margins as Sediment Traps.) *Jl. geol. Soc. Lond.*, **144**, 121–6.

Kornicker, L. S. & H. T. Odum 1958. Characterization of modern and ancient environments by species diversity. *Bull. geol. Soc. Amer.*, **69**, 1599.

Kuenen, P. H. & C. I. Migliorini 1950. Turbidity currents as a cause of graded bedding. *Jl. Geol.*, **58**, 91–127.

Kukal, Z. 1984. *Atlantis in the Light of Modern Research*. Elsevier, Amsterdam, 224 pp. (translated from Czech).

Labeyrie, L. D. & C. Jaendal 1990. Geochemical variation in the oceans, ice and sediments; an introduction. *Palaeogeog., Palaeoclimat. Palaeoecol.* (Global & Planetary Change Sect.), **89**, 1–2.

Lao Tsu 1973. *Tao Te Ching*. (translated from the Chinese by Gia-Fu Feng & J. English), Wildwood House, London, 81 chapters, pages un-numbered.

Larwood, G. P. (edit.) 1988. *Extinction and Survival in the Fossil Record*. Oxford Sci. Publications, 360 pp.

Leakey, M. D. 1987. Introduction – hominid footprints *In* Leakey, M. D. & J. M. Harris. *Laetoli; A Pliocene site in northern Tanzania*. Clarendon Press, Oxford, 490–6.

Lewin, R. 1990. Dinosaur count reveals surprisingly few species. *New Sci.*, No. 1745, 30.

Lewis, K. B. 1971. Slumping on a continental slope inclined at 1°–4°. *Sedimentology*, **16**, 97–110.

Lewis, K. B., R. A. Pickrill & L. Carter 1981. The sand budget of Oriental Bay, Wellington. *New Zealand Ocean. Inst., Fd. Rept.*, No. 17, 27 pp.

Lillie, A. R. 1980. *Strata and Structure in New Zealand*. Tohunga Press, Auckland, NZ, 441 pp.

Loper, D. E., K. McCartney & G. Buzna 1988. A model of correlated episodicity in magnetic-field reversals, climate, and mass extinctions. *Jl. Geol.*, **96**, 1–15.

Lovelock, J. E. 1979. *Gaia: a new look at life on Earth*. Oxford Univ. Press.

Lovelock, J. E. 1983. Gaia as seen through the atmosphere. *In* P. West-broek & E. W. de Jong (edits), *'Biomineralization and Biological Metal Accumulation'*. D. Reidel Publ. Co., 15–25.

Lovelock, J. E. 1988. *The Ages of Gaia*. Norton, New York, 252 pp.

Lovelock, J. E. 1989. The first Leslie Cooper Memorial Lecture given at Plymouth on 10 April 1989: Gaia. *Jl. mar. biol. Ass. UK*, **69**, 746–58.

Lovelock, J. E. & A. J. Watson 1982. The regulation of carbon dioxide and climate. Gaia or geochemistry. *Planetary & Space Sci.*, **30**, 795–802.

Lyell, C. 1830. *Principles of Geology*. vol. I. 1st edn, John Murray, London. (Facsimile reproduction of first edition, with an introduction by M. J. S. Rudwick, 1990, Univ. of Chicago Press, Chicago & London 511 pp.)

Lyell, C. 1832. *Principles of Geology*. vol. II. 1st edn, John Murray, London. (Facsimile reproduction of first edition, 1991, Univ. of Chicago Press, Chicago & London 330 pp.)

McKnight, C. L., S. A. Graham, A. R. Carroll, Q. Gan, D. I. Dilcher, Min Zhao & Yun Hai. Liang 1990. Fluvial sedimentology of an Upper Jurassic petrified forest assemblage, Shishu Formation, Junggar Basin, Xinjiang, China. *Palaeogeog. Palaeoclimat., Palaeoecol.*, **79**, 1–9.

McLaren, D. 1970. Time, life and boundaries. *Jl. Paleont.*, **44**, 801–15.

McRae, L. E. 1990. Palaeomagnetic isochrons, unsteadiness and uniformity of sedimentation in Miocene Intermontane Basin sediments at Salla, eastern Andean Cordillera, Bolivia. *Jl. Geol.*, **98** (4), 479–500.

Malmgren, B. A. & J. P. Kennett 1981. Phyletic gradualism in a Late Cenozoic planktonic foraminiferal lineage, DSDP Site 284, southwest Pacific. *Paleobiol.*, **7**, 230–40.

Manning, S. 1989. A new age for Minoan Crete. *New Sci.*, No. 1651, 60–3.

Martin *et al.* (edits) 1967. *Pleistocene Extinctions*. Yale Univ. Press., New Haven.

Matthews, E. R. 1980. Coastal sediment dynamics, Turakirae Head to East-bourne, Wellington. New Zealand Ocean. Inst., Oceanographic Summary No. 17, 21 pp.

Mayer, L. (edit.) 1987. Catastrophic flooding. Binghampton Symposia in Geomorphology, **18**, 400 pp.

Maynard Smith, J. 1981. Macroevolution. *Nature*, **289**, 13.

Merla, G. 1952. Geologia dell'Appennino Settentrionale. *Boll. Soc. geol. Ital.*, **70** (1951), 95–382.

Merrill, R. D. 1984. *Ophiomorpha* and other non-marine trace fossils from the Eocene Ione Formation, California. *Jl. Paleont.*, **58**, 542–9.

Meyer, L. & D. Nash (edits) 1987. *Catastrophic flooding*. Blackwells, Oxford, 400 pp.

Middleton, L. T., D. K. Elliott & M. Morales 1990. Chapter 10. Coconino Sandstone. *In* S. S. Beus & M. Morales (edits), *Grand Canyon Geology*. Oxford Univ. Press & Mus. North. Arizona Press, 183–202.

Migliorini, C. 1949. I Cunei composti nell'orogenesi. *Boll. geol. Soc. Ital.*, **67** (1948), 29–142.

Migliorini, C. 1952. Composite wedges and orogenic landslips in the Apennines. Congr. Géol. internat., Rept. Sess. XVIII, London, 186–98.

Mitchell, C. & M. Widdowson 1991. A geological map of the southern Deccan Traps, India and its structural implications. *Jl. geol. Soc. Lond.*, **148**, 495–505.

Mudge, D. C. 1978. Stratigraphy and sedimentation of the Lower Inferior Oolite of the Cotswolds. *Jl. geol. Soc. Lond.*, **135**, 611–27.

Napier, W. M. & S. V. M. Clube 1979. A theory of terrestrial catastrophism. *Nature*, **282**, 455–9.

Nelson, C. S., R. M. Briggs & P. J. J. Kamp 1985. Nature and significance of volcanogenic deposits at the Eocene/Oligocene boundary, Hole 593, Challenger Plateau, Tasman Sea. *In* Kennett, J. P. *et al*. Initial Repts. Deep Sea Drilling Project, **90**, 1175–87.

Nelson, R. J. 1837. On the geology of the Bermudas. *Trans. geol. soc. London*, **5** (1), 102–23.

Newell, N. D. 1963. Crises in the history of life. *Scient. Amer.*, **208**, 77–92.

Newell, N. D. 1967. Revolutions in the history of life. *Spec. Paper, geol. Soc. Amer.*, **89**, 63–91.

Noe-Nygaard, N., F. Surlyk & S. Piasecki 1987. Bivalve mass mortality caused by toxic dinoflagellate blooms in a Berriasian–Valanginian lagoon, Bornholm, Denmark. *Palaios*, **2**, 263–73.

Norris, R. M. 1964. Sediments of Chatham Rise. *New Zealand Oceanographic Inst., Mem.*, **26**, /, 9–39.

Officer, C. B., A. Hallam, C. L. Drake & J. D. Devine 1987. Late Cretaceous and paroxysmal Cretaceous/Tertiary Extinctions. *Nature*, **326**, 143–9.

Ostrom, J. H. 1964. A functional analysis of jaw mechanics in the dinosaur *Triceratops. Postilla*, No. 88, 1–35.

Owen, M. R. & M. H. Anders 1988. Evidence from catholuminescence for non-volcanic origin of shocked quartz at the Cretaceous/Tertiary boundary. *Nature*, **334**, 145–7.

Ozawa, T. 1975. Evolution of *Lepidolina multiseptata* (Permian foraminifer) in East Asia. *Mem. Fac. Sci. Kyushu Univ.*, Ser. D Geol., **23**, 117–64.

Pattiaratchi, C. & M. Collins 1987. Mechanisms for linear sandbank formation and maintenance in relation to dynamical and oceanographic observations. *Prog. Oceanog.*, **19**, 117–76.

Pearce, F. 1991. Desert fires cast a shadow over Asia. *New Sci.*, No. 1751, 30–1.

Pedersen, A. K., J. Engell & J. G. Rønsbo 1975. Early Tertiary volcanism in the Skaggerak. New chemical evidence from ash-layers in the Mø Clay of northern Denmark. *Lithos*, **8**, 255–68.

Pedersen, G. K. & F. Surlyk 1983. The Fur Formation, a Late Paleocene ash-bearing diatomite from nothern Denmark. *Bull. geol. Soc. Denmark*, **32**, 43–65.

Pemberton, H. G. & R. W. Frey 1990. Darwin on worms: the advent of experimental neo-ichnology. *Ichnos*, **1**, 65–71.

Penck, A. & E. Bruckner 1909. *Die Alpen im Eiszeitalter*. Leipzig, 267 pp.

Perutz, M. 1986. A new view of Darwinism. *New Scientist*, No. 1528, 36–8.

Pitcher, W. S., D. J. Shearman & D. C. Pugh 1954. The loess at Pegwell Bay, Kent and its associated frost soils. *Geol. Mag.*, **91**, 308–14.

Popper, K. R. 1979. *Objective Knowledge. An evolutionary approach*. 2nd edn Oxford, Clarendon Press, 395 pp.

Posenato, R. 1991. Endemic to Cosmopolitan brachiopods across the P/Tr Boundary in the Southern Alps (Italy). Saito Ho-on Kai spec. Publ. No. 3 (Proc. shallow Tethys Sympos.) Sendai, Japan, 125–39.

Postgate, J. 1988. Gaia gets too big for her boots. *New Sci.*, No. 118, 60.

Putnam, W. C. 1964. '*Geology*'. Oxford Univ. Press, 480 pp.

Ramsay, A. T. S. 1971. A History of the Formation of the Atlantic Ocean. *Brit. Assoc. Advancement Sci.*, 239–49.

Ramsay, A. T. S. 1974. The distribution of calcium carbonate in deep sea sediments. *In*: W. H. Hay (edit.), Studies in Palaeoceanography, *Spec. Publ. Soc. Paleont. & Mineral.*, No. **20**, 58–76.

Ramsay, A. T. S. 1977. Sedimentological clues to palaeoceanography. *In* A. T. S. Ramsay (edit.), '*Oceanic Micropalaeontology*' Academic Press, Orlando, Florida, vol. 2, 1371–453.

Raup, D. M. & J. J. Sepkoski 1982. Mass extinction in the marine fossils. *Science*, **215**, 1501–3.

Raup, D. M. & J. J. Sepkoski 1984. Periodicity of extinctions in the geologic past. *Proc. nat. Acad. Sci.*, **81**, 801–5.

Raup, D. M. & J. J. Sepkoski 1986. Biological extinction in Earth history. *Science*, **231**, 1528–33.

Read, H. H. 1949. A contemplation of time in plutonism. *Quart. Jl. geol. Soc. Lond.*, **105**, 101–56.

Rhoads, D. C. 1966. Depth of burrowing by benthonic marine organisms: a key to nearshore–offshore relationships. *Abstracts geol. Soc. Amer.*, 175.

Rhoads, D. C. 1967. Biogenic reworking of intertidal and subtidal sediments in Barnstable Harbour and Buzzards Bay, Massachusetts. *Jl. Geol.*, **75**, 461.

Rhodes, F. H. T. 1967. Permo-Triassic extinction. *In* Benton, M. & M. A. Whyte (edits), '*The Fossil Record*', Geol. Soc. London, 57–76.

Rhodes, F. H. T. 1987. Darwinian gradualism and its limits: the development of Darwin's views on the rate and pattern of evolutionary change. *Jl. hist. Biol.*, **20** (2), 139–57.

Ridgeway, N. M. 1984. Tsunamis – a natural hazard. *New Zealand DSIR Sci. Inf. Centre*, Pamphlet No. 41, 4 pp.

Rolin, Y., C. Gaillard & M. Roux 1990. Ecologie des pseudobiohermes des Terres Noires jurassiques liés à des paléo-sources sous-marines. Le site oxfordien de Beauvoisin (Drôme, Bassin de Sud-Est France). *Palaeogeog., Palaeoclimat., Palaeoecol.*, **80**, 79–105.

Rose, W. I. & C. A. Chesner 1990. Worldwide dispersal of ash and gases from earth's largest known eruption: Toba, Sumatra. 75 ka. *Palaeogeog., Palaeoclimat., Palaeocol.* (Global & Planetary Change Sect.), **89**, 269–75.

Rosenkrantz, A. & H. W. Rasmussen 1980. South-eastern Sjaelland and Mön, Denmark. *XXI Session internat. geol. Congr.*, Nordern, 16 pp.

Rudwick, M. J. S. 1990. Introduction *In* Charles Lyell, '*Principles of Geology*', volume I. Facsimile reproduction of first edition. Univ. of Chicago Press, Chicago & London, i–lviii.

Scotese, C. R., R. K. Bambach, C. Barton, R. Van Der Voo & A. M. Ziegler 1979. Paleozoic base maps. *Jl. Geol.*, **87** (3), 217–77.

Seilacher, A. 1958. Zur ökologischen Characteristik von Flysch und Molasse. *Eclog. Geol. Helvet.*, **51**, 1062–78.

Seilacher, A. 1967. Bathymetry of trace fossils. *Marine Geol.*, **5**, 413–28.

Sellar, W. C. & R. J. Yeatman 1938. '*1066 and all that*'. Fountain Library/ Methuen, London, 116 pp.

Selley, R. C. 1966. 'The Miocene rocks of the Marada and Jebel Zelten area of central Libya: a study of shoreline sedimentation'. *Petrol. Explor. Soc. Libya*, 30 pp.

Selley, R. C. 1969. Nearshore marine and continental sediments of the Sirte Basin, Libya. *Quart. Jl. geol. Soc. Lond.*, **124**, 419–60.

Shaw, G. B. 1985. '*Pygmalion*'. Penguin Edn., 148 pp.

Simm, R. W. & R. B. Kidd 1984. Submarine debris flow deposits detected by long-range side-scan Sonar 1,000-Kilometers from source. *Geo-Marine Letters*, **3**, 13–6.

Stearn, C. W., R. L. Carroll & T. H. Clark 1979. *Geological Evolution of North America* (3rd edn), John Wiley & Sons, New York, 566 pp.

Stearns, H. T. 1962. Evidence of Lake Bonneville flood along Snake River below King Hill, Idaho. *Bull. geol. Soc. Amer.*, **73**, 385–8.

Stehli, F. G., 1957. Possible Permian climatic zonation and its implications. *Amer. Jl. Sci.*, **255**, 607–18.

Stoneley, H. M. 1956. *Hiltonia*, a new plant genus from the Upper Permian of England. *Ann. Mag. nat. Hist.*, Ser. **12**, 9, 713–20.

Stoneley, H. M. 1958. The Upper Permian Flora of England. *Bull. brit. Mus. (nat. Hist.)*, Geology, **3** (9), 295–337.

Stothers, R. B. 1988. Structure of Oort's comet cloud inferred from terrestrial impact craters. *The Observatory*, **108**, No. 1082, 1–9.

Strong, C. P. 1984. Cretaceous-Tertiary boundary, Mid-Waipara River section, North Canterbury, New Zealand. *New Zealand Jl. Geol.*, **27**, 231–4.

Surlyk, F. 1980. The Cretaceous-Tertiary Boundary event. *Nature*, **285**, 187–8.

Surlyk, F. & M. Bagge-Johansen 1984. End-Cretaceous Brachiopod Extinctions in the Chalk of Denmark. *Amer. Assoc. Adv. Sci.*, **223**, 1174–7

Surlyk, F. & W. K. Christensen 1974. Epifaunal zonation on an Upper Cretaceous rocky coast. *Geology*, **2** (II), 529–34.

Swain, T. 1974. Cold-blooded murder in the Cretaceous. *Spectrum*, No. 120, 10–12.

Talbot, M. R. & K. Kelts (edits) 1989. The Phanerozoic record of lacustrine basins and their environmental signals. *Palaeogeog., Palaeoclimat., Palaeoecol.*, **70**, 304 pp.

Taylor, D. W. 1991. Paleobiogeographic relationships of Andean angiosperms of Cretaceous to Pliocene age. *Palaeogeog., Palaeoclimat., Palaeoecol.*, **88**, 69–84.

Teichert, C. 1986. Times of crisis in the evolution of the Cephalopoda. *Paläont. Zeitschr.*, **60**, 227–43.

Todd, J. A. 1991. A forest-litter animal community from the Upper Carboniferous: notes on the association of animal body fossils with plants and lithology in the Westphalian D Coal Measures at Writhlington, Avon. *Proc. Geol. Ass.*, **102** (3), 179–84.

Trueman, A. E. 1922. The use of *Gryphaea* in the correlation of the Lower Lias, *Geol. Mag.*, **59**, 256–68.

Trueman, A. E. & J. Weir 1946. A Monograph of British Carboniferous non-marine Lamellibranchia. Part I, *Palaeontogr. Soc.*, **99**, 1–16.

Trümpy, R. 1980. Trends, discoveries, failures. Address presented at the Opening Ceremony of the 26th International Geological Congress, Paris, 5 pp.

Tudge, C. 1989. Evolution and the end of innocence, *New Sci.*, No. 1750, 26–30.

Tudge, C. 1991. A wild time at the zoo. *New Sci.*, No. 1750, 26–30.

Urey, H. C. 1973. Cometary collisions and geological periods. *Nature*, **242**, 32–3.

Veevers, J. J. (edit.) 1984. *Phanerozoic earth history of Australia*. Clarendon Press, Oxford, xv + 418 pp.

Verchnur, G. L. 1991. The End of Civilization? *Astronomy*, **19** (9), 51–4.

'Verifier' 1877. *Scepticism in Geology and the reasons for it*. John Murray, London, 126 pp.

Vinogradov, A. P. (edit.) 1961. Atlas of lithological-palaeogeographical maps of the Russian platform and its geosynclinal framing [sic], pt. II, Mesozoic and Cenozoic, Sheets 48–95, Moscow. (in Russian with some English captions).

Waitt, H. B. 1985. Case for periodic colossal jökulhaups from Pleistocene Lake Missoula. *Bull. geol. Soc. Amer.*, **96**, 1271–86.

Walker, J. C. G. 1977. *Evolution of the atmosphere.* Macmillan (New York), 318 pp.

Walker, J. C. G. 1982. Climatic factors on the Archaean Earth. *Palaeogeog., Paleoclimat., Paleoecol.*, **40**, 1–11.

Wallace, P. 1969. The sedimentology and palaeoecolgy of the Devonian of the Ferques inlier. northern France. *Quart. Jl. geol. Soc. Lond.*, **125**, 83–124.

Wallace, P. 1972. The geology of the Palaeozoic rocks of the south-western part of the Cantabrian Cordillera, north Spain. *Proc. Geol. Ass.*, **83**, 57–73.

Walley, C. D. 1988. A braided strike-slip model for the northern continuation of the Dead Sea Fault and its implications for Levantine tectonics. *Tectonophysics*, **145**, 63–72.

Wanless, H. R. 1960. Evidences of multiple Late Paleozoic glaciation in Australia. *Internat. geol. Congr.*, XXI Session, Norden, Pt. XII, 104–10.

Wanless, H. R., L. P. Tedesco & K. M. Tyrrell 1988. Production of subtidal and surficial tempestites by hurricane Kate, Caioes Platform, British West Indies. *Jl. sediment. Petrology*, **58**, 739–50.

Webster, G. D., M. J. Pankratz Kuhns & G. L. Waggoner 1982. Late Cenozoic gravels in Hell's Canyon and the Lewiston Basin, Washington and Idaho, *In* B. Bonishen & R. M. Breckenridge (edits), 'Cenozoic geology of Idaho'. *Bull. Idaho Bur. Mines & Geol.*, **26**, 669–83.

Wendt, H. 1970. *Before the Deluge.* Paladin, London, 428 pp.

Wezel, F-C. 1981. Reply (to Alvarez & Lowrie). *Nature*, **294**, 248.

Wezel, F-C. 1988. Earth structural patterns and rhythmic tectonism. *Tectonophysics*, **146**, 1–45.

Wezel, F-C. 1992. Global change in earth history: a personal point of view. *Terra Nova,* In the press.

Wezel, F-C., S. Vanucci & R. Vanucci 1981. Découverte de divers niveaux riches en iridium dans le 'Scaglia rossa' et la 'Scaglia blanca' de l'Appenin d'Ombrie, Marches (Italie). Comptes Rendues de l'Acad. Sci. de Paris II, **293**, 837–44.

White, Gilbert. 1788 (original edition; my edition 1905). *The Natural History and Antiquities of Selborne.* Routledge & Sons, 475 pp.

Whittard, W. F. 1932. The stratigraphy of the Valentian rocks of Shropshire. The Longmynd-Shelve and Breidden outcrops, *Quart. Jl. geol. Soc. Lond.*, **88**, 859–902.

Williamson, P. G. 1981a. Palaeontological documentation of speciation in Cenozoic molluscs from Turkana Basin. *Nature*, **293**, 437–43.

Williamson, P. G. 1981b. Morphological stasis and developmental constraint: real problems for neo-Darwinism. *Nature*, **294**, 214–5.

Williamson, P. G. 1982a. Molluscan biostratigraphy of the Koobi Fora hominid-bearing deposits. *Nature*, **295**, 140–2.

Williamson, P. G. 1982b. Reply: Punctuationism and Darwinism reconciled? The Lake Turkana mollusc sequence. *Nature*, **296**, 4–5.

Woodward, F. I. 1987. Stomatal numbers are sensitive to increases in CO_2 from pre-industrial levels. *Nature*, **327**, 617–8.

Wright, V. P. & S. D. Vanstone 1991. Assessing the carbon dioxide content of ancient atmospheres using palaeo-calcretes: theoretical and empirical constraints [*sic*]. *Jl. geol. Soc. Lond.*, **148**, 945–7.

Wu, X-T. 1982. Storm-generated depositional types and associated trace-fossils in Lower Carboniferous shallow-marine carbonates of Three Cliffs Bay and Ogmore-by-Sea, South Wales. *Palaeogeog., Palaeoclimat., Palaeoecol.*, **39**, 187–262.

Zezina, O. N. 1985. *Recent brachiopods and problems of the bathyal zone of the oceans*. Izdatel. Nauka, Moscow, 247 pp. (in Russian).

Zoller, W. H., J. R. Parrington & J. M. Phelan-Kotra 1983. Iridium enrichment in airborne particles from Kilauea Volcano: January 1983. *Science*, **222**, 1118–21.

Index

N.b. Adjectives are listed under the relevant nouns, e.g. 'British' under 'Britain'.

218 Index